Developing
Essential Understanding
of
Addition and Subtraction
for Teaching Mathematics *in*
Prekindergarten–Grade 2

Janet H. Caldwell
Rowan University
Glassboro, New Jersey

Karen Karp
University of Louisville
Louisville, Kentucky

Jennifer M. Bay-Williams
University of Louisville
Louisville, Kentucky

Ed Rathmell
Volume Editor
University of Northern Iowa
Cedar Falls, Iowa

Rose Mary Zbiek
Series Editor
The Pennsylvania State University
University Park, Pennsylvania

NATIONAL COUNCIL OF
TEACHERS OF MATHEMATICS

Library of Congress Cataloging-in-Publication Data

Developing essential understanding of addition and subtraction for teaching mathematics in prekindergarten-grade 2 / Janet H. Caldwell ... [et al.].
    p. cm.
  ISBN 978-0-87353-664-6
  1. Addition--Study and teaching (Preschool) 2. Addition--Study and teaching (Primary) 3. Subtraction--Study and teaching (Preschool) 4. Subtraction--Study and teaching (Primary) I. Caldwell, Janet H.
    QA135.6.D486 2010
    372.7'2--dc22
                        2010036661

The National Council of Teachers of Mathematics is a public voice of mathematics education, supporting teachers to ensure equitable mathematics learning of the highest quality for all students through vision, leadership, professional development, and research.

Printed in the United States of America

Contents

Foreword

Teaching mathematics in prekindergarten–grade 12 requires a special understanding of mathematics. Effective teachers of mathematics think about and beyond the content that they teach, seeking explanations and making connections to other topics, both inside and outside mathematics. Students meet curriculum and achievement expectations when they work with teachers who know what mathematics is important for each topic that they teach.

The National Council of Teachers of Mathematics (NCTM) presents the Essential Understanding Series in tandem with a call to focus the school mathematics curriculum in the spirit of *Curriculum Focal Points for Prekindergarten through Grade 8 Mathematics: A Quest for Coherence*, published in 2006, and *Focus in High School Mathematics: Reasoning and Sense Making*, released in 2009. The Essential Understanding books are a resource for individual teachers and groups of colleagues interested in engaging in mathematical thinking to enrich and extend their own knowledge of particular mathematics topics in ways that benefit their work with students. The topic of each book is an area of mathematics that is difficult for students to learn, challenging to teach, and critical for students' success as learners and in their future lives and careers.

Drawing on their experiences as teachers, researchers, and mathematicians, the authors have identified the big ideas that are at the heart of each book's topic. A set of essential understandings—mathematical points that capture the essence of the topic—fleshes out each big idea. Taken collectively, the big ideas and essential understandings give a view of a mathematics that is focused, connected, and useful to teachers. Links to topics that students encounter earlier and later in school mathematics and to instruction and assessment practices illustrate the relevance and importance of a teacher's essential understanding of mathematics.

On behalf of the Board of Directors, I offer sincere thanks and appreciation to everyone who has helped to make this series possible. I extend special thanks to Rose Mary Zbiek for her leadership as series editor. I join the Essential Understanding project team in welcoming you to these books and in wishing you many years of continued enjoyment of learning and teaching mathematics.

Henry Kepner
President, 2008–2010
National Council of Teachers of Mathematics

Preface

From prekindergarten through grade 12, the school mathematics curriculum includes important topics that are pivotal in students' development. Students who understand these ideas cross smoothly into new mathematical terrain and continue moving forward with assurance.

However, many of these topics have traditionally been challenging to teach as well as learn, and they often prove to be barriers rather than gateways to students' progress. Students who fail to get a solid grounding in them frequently lose momentum and struggle in subsequent work in mathematics and related disciplines.

The Essential Understanding Series identifies such topics at all levels. Teachers who engage students in these topics play critical roles in students' mathematical achievement. Each volume in the series invites teachers who aim to be not just proficient but outstanding in the classroom—teachers like you—to enrich their understanding of one or more of these topics to ensure students' continued development in mathematics.

How much do you need to know?

To teach these challenging topics effectively, you must draw on a mathematical understanding that is both broad and deep. The challenge is to know considerably more about the topic than you expect your students to know and learn.

Why does your knowledge need to be so extensive? Why must it go above and beyond what you need to teach and your students need to learn? The answer to this question has many parts.

To plan successful learning experiences, you need to understand different models and representations and, in some cases, emerging technologies as you evaluate curriculum materials and create lessons. As you choose and implement learning tasks, you need to know what to emphasize and why those ideas are mathematically important.

While engaging your students in lessons, you must anticipate their perplexities, help them avoid known pitfalls, and recognize and dispel misconceptions. You need to capitalize on unexpected classroom opportunities to make connections among mathematical ideas. If assessment shows that students have not understood the material adequately, you need to know how to address weaknesses that you have identified in their understanding. Your understanding must be sufficiently versatile to allow you to represent the mathematics in different ways to students who don't understand it the first time.

In addition, you need to know where the topic fits in the full span of the mathematics curriculum. You must understand where your students are coming from in their thinking and where they are heading mathematically in the months and years to come.

Accomplishing these tasks in mathematically sound ways is a tall order. A rich understanding of the mathematics supports the varied work of teaching as you guide your students and keep their learning on track.

How can the Essential Understanding Series help?

The Essential Understanding books offer you an opportunity to delve into the mathematics that you teach and reinforce your content knowledge. They do not include materials for you to use directly with your students, nor do they discuss classroom management, teaching styles, or assessment techniques. Instead, these books focus squarely on issues of mathematical content—the ideas and understanding that you must bring to your preparation, in-class instruction, one-on-one interactions with students, and assessment.

How do the authors approach the topics?

For each topic, the authors identify "big ideas" and "essential understandings." The big ideas are mathematical statements of overarching concepts that are central to a mathematical topic and link numerous smaller mathematical ideas into coherent wholes. The books call the smaller, more concrete ideas that are associated with each big idea *essential understandings*. They capture aspects of the corresponding big idea and provide evidence of its richness.

The big ideas have tremendous value in mathematics. You can gain an appreciation of the power and worth of these densely packed statements through persistent work with the interrelated essential understandings. Grasping these multiple smaller concepts and through them gaining access to the big ideas can greatly increase your intellectual assets and classroom possibilities.

In your work with mathematical ideas in your role as a teacher, you have probably observed that the essential understandings are often at the heart of the understanding that you need for presenting one of these challenging topics to students. Knowing these ideas very well is critical because they are the mathematical pieces that connect to form each big idea.

How are the books organized?

Every book in the Essential Understanding Series has the same structure:

- The introduction gives an overview, explaining the reasons for the selection of the particular topic and highlighting some of the differences between what teachers and students need to know about it.

- Chapter 1 is the heart of the book, identifying and examining the big ideas and related essential understandings.

Big ideas and essential understandings are identified by icons in the books.

marks a big idea, and

marks an essential understanding.

- Chapter 2 reconsiders the ideas discussed in chapter 1 in light of their connections with mathematical ideas within the grade band and with other mathematics that the students have encountered earlier or will encounter later in their study of mathematics.

- Chapter 3 wraps up the discussion by considering the challenges that students often face in grasping the necessary concepts related to the topic under discussion. It analyzes the development of their thinking and offers guidance for presenting ideas to them and assessing their understanding.

The discussion of big ideas and essential understandings in chapter 1 is interspersed with questions labeled "Reflect." It is important to pause in your reading to think about these on your own or discuss them with your colleagues. By engaging with the material in this way, you can make the experience of reading the book participatory, interactive, and dynamic.

Reflect questions can also serve as topics of conversation among local groups of teachers or teachers connected electronically in school districts or even between states. Thus, the Reflect items can extend the possibilities for using the books as tools for formal or informal experiences for in-service and preservice teachers, individually or in groups, in or beyond college or university classes.

A new perspective

The Essential Understanding Series thus is intended to support you in gaining a deep and broad understanding of mathematics that can benefit your students in many ways. Considering connections between the mathematics under discussion and other mathematics that students encounter earlier and later in the curriculum gives the books unusual depth as well as insight into vertical articulation in school mathematics.

The series appears against the backdrop of *Principles and Standards for School Mathematics* (NCTM 2000), *Curriculum Focal Points for Prekindergarten through Grade 8 Mathematics: A Quest for Coherence* (NCTM 2006), *Focus in High School Mathematics: Reasoning and Sense Making* (NCTM 2009), and the Navigations Series (NCTM 2001–2009). The new books play an important role, supporting the work of these publications by offering content-based professional development.

The other publications, in turn, can flesh out and enrich the new books. After reading this book, for example, you might select hands-on, Standards-based activities from the Navigations books for your students to use to gain insights into the topics that the Essential Understanding books discuss. If you are teaching students in prekindergarten through grade 8, you might apply your deeper understanding as you present material related to the three focal points that *Curriculum Focal Points* identifies for instruction at your students' level. Or if you are teaching students in grades 9–12, you might use your understanding to enrich the ways in which you can engage students in mathematical reasoning and sense making as presented in *Focus in High School Mathematics.*

An enriched understanding can give you a fresh perspective and infuse new energy into your teaching. We hope that the understanding that you acquire from reading the book will support your efforts as you help your students grasp the ideas that will ensure their mathematical success.

We appreciate the thoughtful comments and suggestions offered by the following individuals who reviewed an earlier version of this book: Linda Coutts, Terry Crites, Corey Drake, Walter Seaman, and Erica Guenther Steele.

Introduction

This book focuses on ideas about addition and subtraction, including connections to many mathematical ideas that are common in primary school curricula. These are ideas that you must understand thoroughly and be able to use flexibly to be highly effective in your teaching of mathematics in prekindergarten through grade 2. The authors have assumed that you have had a variety of mathematics experiences that have motivated you to delve into—and move beyond—the mathematics that you expect your students to learn.

The book is designed to engage you as a learner of mathematics, focusing on content and curricular considerations that will better equip you to plan and implement lessons and assess your students' learning in ways that reflect the full complexity of addition and subtraction. A deep, rich understanding of these operations and their properties will enable you to communicate their influence and scope to your students, showing them how these ideas relate to the mathematics that they have already encountered—and will continue to encounter—throughout their school mathematics experiences.

The understanding of addition and subtraction that you gain from this focused study thus supports the vision of *Principles and Standards for School Mathematics* (NCTM 2000): "Imagine a classroom, a school, or a school district where all students have access to high-quality, engaging mathematics instruction" (p. 3). This vision depends on classroom teachers who "are continually growing as professionals" (p. 3) and routinely engage their students in meaningful experiences that help them learn mathematics with understanding.

Why Addition and Subtraction?

Like the topics of all the volumes in the Essential Understanding Series, addition and subtraction compose a major area of school mathematics that is crucial for students to learn but challenging for teachers to teach. Students in prekindergarten through grade 2 need to understand these operations well if they are to succeed in these grades and in their subsequent mathematics experiences. Learners often struggle with ideas about addition and subtraction. What is the relationship between addition and subtraction? How do problem solvers know whether an algorithm will always work? Many students mistakenly believe that these operations have identical properties, and they often compute as though there were a universal associative property. The importance of addition and subtraction and the challenge of understanding them well make them essential for teachers of prekindergarten through grade 2 to understand in great depth.

Beyond having a solid understanding of addition and subtraction—operations that are complex in and of themselves—you must also know how concepts and properties of addition and subtraction relate to other mathematical ideas that your students will encounter in the lesson at hand, the current school year, and beyond. Addition and subtraction are essential to understand deeply because they are foundational to many other mathematical concepts within the prekindergarten–grade 2 curriculum and beyond. Having a rich understanding of addition and subtraction will influence your instructional decisions in such areas as the following:

- ✓ Selecting tasks for a lesson

- ✓ Developing key questions to pose to students

- ✓ Choosing materials for exploring relevant mathematics

- ✓ Ordering topics and ideas over time

- ✓ Assessing the quality of students' work

- ✓ Devising ways to challenge and support students' thinking

This checklist, although not exhaustive, can serve as a self-assessment or guide as you read. Asking yourself how the particular mathematical idea under discussion can influence your work with students will maximize the impact of this book and provide a clear answer to the question, "Why develop essential understanding of addition and subtraction?"

Understanding Addition and Subtraction

Teachers teach mathematics because they want others to understand it in ways that will contribute to success and satisfaction in school, work, and life. Helping your young students develop a robust and lasting understanding of addition and subtraction requires that you understand this mathematics deeply. But what does this mean?

It is easy to think that understanding an area of mathematics, such as addition and subtraction, means only knowing certain facts, being able to use basic techniques to solve particular types of problems, and mastering relevant vocabulary. For example, at the primary level, you are expected to know such facts as "addition of whole numbers is a commutative operation." You are expected to be skillful in solving problems that involve multiple steps. Your mathematical vocabulary is assumed to include such terms as *sum*, *difference*, *inverse*, *fact family*, and *whole number.*

Obviously, facts, vocabulary, and techniques for solving certain types of problems are not all that you are expected to know about addition and subtraction. For example, in your ongoing work with students, you have undoubtedly discovered that you need to distinguish among different addition and subtraction strategies, such as knowing the difference between "counting all" and "counting on."

It is also tempting to focus only on a long list of mathematical ideas that all teachers of mathematics in prekindergarten through grade 2 are expected to know and teach about addition and subtraction. Often state mathematics standards are examples of such lists. However important the individual items might be, these lists cannot capture the essence of a dynamic understanding of the topic. Understanding these operations deeply requires that you not only know important mathematical ideas but also recognize how these ideas relate to one another. For example, knowing the properties of addition is important, but so is realizing how they are applied in algorithms and generalizations that students generate (for example, that + 9 is the same result as + 10 − 1). Your dynamic understanding continues to grow with experience and as a result of opportunities to embrace new ideas and find new connections among familiar ones.

Furthermore, your understanding of addition and subtraction should transcend the content intended for your students. In some cases, it is clear why you need to know more mathematics than your students. For instance, your understanding of how addition and subtraction properties connect with multiplication and division properties—mathematics that your students will encounter later—can influence your choice of instructional tasks and the terminology that you use. You might have your students explore when commutativity and associativity hold (with addition, but not subtraction) and why this is the case—foundations that they can later build on and extend by exploring with multiplication and division.

Other differences between the understanding that you need to have and the understanding that you expect your students to acquire are less obvious, but your experiences in the classroom have undoubtedly made you aware of them at some level. For example, students commonly ask, "Can I do it this way?" or "Does this always work?" or "Is this the same as that?" They naturally attempt to make generalizations or connections. In such instances, how many times have you been grateful to have an understanding of addition and subtraction that enables you to recognize the merit in a student's unanticipated mathematical question or claim? How many other times have you wondered whether you could be missing an opportunity because of a gap in your knowledge?

As you have almost certainly discovered, knowing and being able to do familiar mathematics are not enough when you're in the classroom. You also need to be able to identify and justify or refute novel claims—or, better yet, assist students in thinking about these claims in guided conversations. Such claims and justifications might draw on ideas or techniques that are beyond the mathematical experiences of your students and current curricular expectations for them. For example, you may need to be able to refute the often-asserted, erroneous claim that the sum of any two whole numbers is greater than either of the two addends (take, for example, 6 + 0), or you may need to explain to a student why order matters in subtracting whole numbers but not in adding them.

The Big Ideas and Essential Understandings

Thinking about the many particular ideas that are part of a rich understanding of addition and subtraction can be an overwhelming task. Articulating all of those mathematical ideas and their connections would require many resources. To choose which ideas to include in this book, the authors considered a critical question: What is *essential* for teachers of mathematics in prekindergarten through grade 2 to know about addition and subtraction to be effective in the classroom? To answer this question, the authors drew on a variety of resources, including personal experiences, the expertise of colleagues in mathematics and mathematics education, the wisdom of classroom teachers, and the reactions of reviewers and professional development providers, as well as ideas from curricular materials and research on mathematics learning and teaching.

As a result, the mathematical content of this book focuses on essential ideas for teachers about addition and subtraction. In particular, chapter 1 is organized around two big ideas related to this important area of mathematics. Each of these big ideas is supported by more specific mathematical ideas called essential understandings.

Benefits for Teaching, Learning, and Assessing

Understanding addition and subtraction can help you implement the Teaching Principle enunciated in *Principles and Standards for School Mathematics*. This Principle sets a high standard for instruction: "Effective mathematics teaching requires understanding what students know and need to learn and then challenging and supporting them to learn it well" (NCTM 2000, p. 16). As in teaching about other critical topics in mathematics, teaching about addition and subtraction requires knowledge that goes "beyond what most teachers experience in standard preservice mathematics courses" (p. 17).

Chapter 1 comes into play at this point, offering an overview of addition and subtraction that is intended to be more focused and comprehensive than many discussions of the topic that you are likely to have encountered. This chapter enumerates, expands on, and gives examples of the big ideas and essential understandings related to the two operations and their properties, with the goal of supplementing or reinforcing your understanding. Thus, chapter 1 aims to prepare you to implement the Teaching Principle fully as you provide the support and challenge that your students need for robust learning about addition and subtraction.

Consolidating your understanding in this way also prepares you to implement the Learning Principle outlined in *Principles and Standards*: "Students must learn mathematics with understanding, actively building new knowledge from experience and prior knowledge" (NCTM 2000, p. 20). To support your efforts to help your students learn about addition and subtraction in this way, chapter 2 extends your understanding of these operations by identifying specific ways in which the big ideas and essential understandings connect with mathematics that students typically encounter earlier or later in school. This chapter supports the Learning Principle by emphasizing longitudinal connections. For example, as their mathematical experiences expand, students gradually develop an understanding of the connections between addition of whole numbers and addition of decimals, and they can use various representations as needed.

The understandings that chapters 1 and 2 convey can strengthen another critical area of teaching. Chapter 3 addresses this area, building on the first two chapters to show how an understanding of addition and subtraction can help you select and develop appropriate tasks, techniques, and tools for assessing your students' understanding of addition and subtraction. An ownership of the big ideas and essential understandings related to addition and subtraction, reinforced by an understanding of students' past and future experiences with these ideas, can help you ensure that assessment in your classroom supports the learning of significant mathematics.

Such assessment satisfies the first requirement of the Assessment Principle set out in *Principles and Standards* (NCTM 2000): "Assessment should support the learning of important mathematics and furnish useful information to both teachers and students" (p. 22). An understanding of addition and subtraction can also help you satisfy the second requirement of the Assessment Principle, by enabling you to develop assessment tasks that give you specific information about what your students are thinking and what they understand. For example, the following task challenges students to think about the meaning of the equals sign and the meaning of an equation in a task like the following:

What number do you think belongs in the box?

$$8 + 4 = \Box + 5$$

This kind of task taps children's thinking and provides information that can point to better ways to support their learning.

Ready to Begin

The introduction has painted the background, preparing you for the big ideas and associated essential understandings that you will encounter in chapter 1. Reading the chapters in the order in which they appear can be a very useful way to approach the book. Read chapter 1 in more than one sitting, allowing time for thoughtful examination of the Reflect questions embedded in the text. Absorb the ideas—both big ideas and essential understandings—related to addition and subtraction. Appreciate the connections among these ideas. Discuss them with colleagues. Carry your newfound or reinforced understanding to chapter 2, which guides you in seeing how the ideas related to addition and subtraction are connected to the mathematics that your students have encountered earlier or will encounter later in school. Then read about teaching, learning, and assessment issues in chapter 3 and test these ideas in your classroom.

Alternatively, you may want to take a look at chapter 3 before engaging with the mathematical ideas in chapters 1 and 2. Reading the book in this way, with the challenges of teaching, learning, and assessment clearly in mind, along with possible approaches to them, will enable you to gain a different perspective on the material in the earlier chapters. Use the checklist provided earlier as a guide for thinking about new knowledge you have acquired across these dimensions of teaching.

No matter how you read the book, let it serve as a tool to expand your understanding, application, and enjoyment of addition and subtraction.

Addition and Subtraction: The Big Ideas and Essential Understandings

Young children begin learning mathematics before they enter school. They learn to count, and they can solve simple problems by counting. In the primary grades, mathematics instruction focuses on the development of number sense, understanding of numerical operations, and fluency in performing computations. *Principles and Standards for School Mathematics* (National Council of Teachers of Mathematics [NCTM] 2000) describes the development of these skills and concepts, asserting that by the end of grade 2, students should—

- understand numbers, ways of representing numbers, relationships among numbers, and number systems;
- understand meanings of operations and how they relate to one another;
- compute fluently and make reasonable estimates. (NCTM 2000, p. 78)

Curriculum Focal Points for Prekindergarten through Grade 8 Mathematics: A Quest for Coherence (NCTM 2006) continues to emphasize the importance of developing both conceptual understanding and procedural understanding of addition and subtraction. Building on *Principles and Standards, Curriculum Focal Points* recommends that instruction focus on developing this understanding throughout the early grades. Activities in kindergarten should center on joining and separating sets:

> Children use numbers, including written numerals, to represent quantities and to solve quantitative problems, such as . . . modeling simple joining and separating situations with objects. They choose, combine, and apply effective strategies for answering quantitative questions, including quickly recognizing the number in a small set, counting and producing sets of given sizes, counting the number in combined sets, and counting backward. (NCTM 2006, p. 12)

In grade 1, instruction should focus on developing students' understanding of addition and subtraction as well as related facts and strategies associated with these operations:

> Children develop strategies for adding and subtracting whole numbers on the basis of their earlier work with small numbers. They use a variety of models, including discrete objects, length-based models (e.g., lengths of connecting cubes), and number lines, to model "part-whole," "adding to," "taking away from," and "comparing" situations to develop an understanding of the meanings of addition and subtraction and strategies to solve such arithmetic problems. Children understand the connections between counting and the operations of addition and subtraction (e.g., adding two is the same as "counting on" two). They use properties of addition (commutativity and associativity) to add whole numbers, and they create and use increasingly sophisticated strategies based on these properties (e.g., "making tens") to solve addition and subtraction problems involving basic facts. By comparing a variety of solution strategies, children relate addition and subtraction as inverse operations. (NCTM 2006, p. 13)

In grade 2, the instructional focus should shift to helping students develop quick recall of addition and related subtraction facts, as well as fluency with multi-digit addition and subtraction:

> Children use their understanding of addition to develop quick recall of basic addition facts and related subtraction facts. They solve arithmetic problems by applying their understanding of models of addition and subtraction (such as combining or separating sets or using number lines), relationships and properties of number (such as place value), and properties of addition (commutativity and associativity). Children develop, discuss, and use efficient, accurate, and generalizable methods to add and subtract multidigit whole numbers. They select and apply appropriate methods to estimate sums and differences or calculate them mentally, depending on the context and numbers involved. They develop fluency with efficient procedures, including standard algorithms, for adding and subtracting whole numbers, understand why the procedures work (on the basis of place value and properties of operations), and use them to solve problems. (NCTM 2006, p. 14)

This chapter discusses in detail the mathematical concepts that these publications outline. In addition, it relates various mathematical processes to these concepts by exploring—

- situations for which addition and subtraction can be used to solve problems;

- ways to represent addition and subtraction;

- ways to reason with addition and subtraction; and

- connections and relationships among these and other mathematical topics.

It also examines numerical relationships that arise from studying multiple representations and the reasoning required for the meaningful use and understanding of computational algorithms, written and mental, standard and nonstandard. The representations have been chosen primarily for their usefulness in illustrating the mathematical concepts. Most of the early examples use counters, since these constitute the most elementary representation, but later discussions involve the use of other representations for addition and subtraction, such as the number line, a hundreds chart, and base-ten place-value blocks.

"Unpacking" ideas related to addition and subtraction is a critical step in establishing deeper understanding. To someone without training as a teacher, these ideas might appear to be simple to teach. But those who teach young students are aware of the subtleties and complexities of the ideas themselves and the challenges of presenting them clearly and coherently in the classroom. Teachers of young students also have an idea of the overarching importance of addition and its inverse operation, subtraction:

Overarching idea: Addition and its inversely related operation, subtraction, are powerful foundational concepts in mathematics, with applications to many problem situations and connections to many other topics. Addition determines the whole in terms of the parts, and subtraction determines a missing part.

This overarching idea anchors teachers' understanding and their instruction. It incorporates two big ideas about addition and subtraction that are crucial to understand. The first relates to when to use each operation, and the second deals with how to get answers efficiently and accurately. Each of these two big ideas involves several smaller, more specific essential understandings.

These big ideas and essential understandings are identified here as a group to give you a quick overview and for your convenience in referring back to them later. Read through them now, but do not think that you must absorb them fully at this point. The chapter will discuss each one in turn in detail.

Big Idea 1. Addition and subtraction are used to represent and solve many different kinds of problems.

> **Essential Understanding 1a.** Addition and subtraction of whole numbers are based on sequential counting with whole numbers.
>
> **Essential Understanding 1b.** Subtraction has an inverse relationship with addition.
>
> **Essential Understanding 1c.** Many different problem situations can be represented by part-part-whole relationships and addition or subtraction.
>
> **Essential Understanding 1d.** Part-part-whole relationships can be expressed by using number sentences like $a + b = c$ or $c - b = a$, where a and b are the parts and c is the whole.
>
> **Essential Understanding 1e.** The context of a problem situation and its interpretation can lead to different representations.

Big Idea 2. The mathematical foundations for understanding computational procedures for addition and subtraction of whole numbers are the properties of addition and place value.

> **Essential Understanding 2a.** The commutative and associative properties for addition of whole numbers allow computations to be performed flexibly.
>
> **Essential Understanding 2b.** Subtraction is not commutative or associative for whole numbers.
>
> **Essential Understanding 2c.** Place-value concepts provide a convenient way to compose and decompose numbers to facilitate addition and subtraction computations.
>
> **Essential Understanding 2d.** Properties of addition are central in justifying the correctness of computational algorithms.

Representing and Solving Problems: Big Idea 1

Big Idea 1. *Addition and subtraction are used to represent and solve many different kinds of problems.*

Many different types of problems can be represented by addition or subtraction. It is important to learn how to recognize these situations and represent them symbolically, building on counting with whole numbers. By understanding these situations and their representations well, teachers can provide students with many different examples of addition and subtraction problems. The discussion below of Big Idea 1 presents and examines fifteen examples that illustrate situations that can be represented by addition or subtraction.

Building on sequential counting

Essential Understanding 1a. *Addition and subtraction of whole numbers are based on sequential counting with whole numbers.*

Situations that can be represented by addition or subtraction can be considered as basic applications of counting forward or back. Even very young children can solve simple addition and subtraction story problems by counting concrete objects (e.g., Starkey and Gelman 1982; Carpenter and Moser 1983). They establish a one-to-one correspondence by moving, touching, or pointing to each object that they are counting as they say the corresponding number words. The following two examples demonstrate how counting relates to addition and subtraction situations.

Example 1 lends itself to a number of simple counting strategies:

> **Example 1:** Max has 2 apples. He picks 5 more. How many apples does Max have now?

For an extended discussion of counting strategies and number ideas, see *Developing Essential Understanding of Number and Numeration for Teaching Mathematics in Pre-kindergarten–Grade 2* (Dougherty et al. 2010).

This problem can be represented with concrete objects by first placing 2 counters (the quantity that Max starts with) and then placing a second group of 5 more counters. Figure 1.1 illustrates the two groups of counters.

Fig. 1.1. Counters representing 2 and 5

A variety of counting strategies might be used to find the total number of counters:

- *Count all:* Count each of the counters: 1, 2 [pause] 3, 4, 5, 6, 7.

- *Count on from the first number:* A more efficient way to find the total is to count on, beginning with the first quantity given in the problem (in this case, 2): 2 [pause], 3, 4, 5, 6, 7.

- *Count on from the larger number:* A still more efficient way to find the total is to count on, beginning with the larger number (5, in this case) and counting on the smaller number (2): 5, [pause] 6, 7.

Reflect 1.1 explores possible ways to use counters with these strategies.

Reflect 1.1

What counters might a child point to as she uses each of the counting strategies shown above?

Does a child need all of the counters for counting on?

Each of the "counting on" strategies is more efficient when problem solvers recognize the first number that they use without counting. This process reduces the difficulty of many tasks and is frequently useful in playing games, counting coins, or other simple everyday tasks. Recognizing patterns on number cubes and dominoes, such as in figure 1.2 is particularly helpful.

Fig. 1.2. Number cubes and dominoes

Relating numbers to the benchmark quantities 5 and 10 helps students see the relative sizes of numbers and can therefore support their transition from counting to later work in addition and subtraction. In the five-frame on the left in figure 1.3, we not only recognize the three counters without counting, but we also note without counting that there are two empty spaces, so 3 is 2 less than 5, or $3 + 2 = 5$, or $5 - 2 = 3$. In the ten-frame on the right, we see that 6 is 1 more than 5, or $6 = 5 + 1$, and that 6 is 4 less than 10, or $6 + 4 = 10$, or $6 = 10 - 4$.

Fig. 1.3. Five-frame and ten-frame representing 3 and 6, respectively

Working with dot patterns, whether on a ten-frame or a number cube, can help students recognize the number of objects without counting. Some representations are more useful for building recognition of multiples—especially doubles. The arrangement of six dots on a number cube is similar to the arrangement of counters on the ten-frame shown in figure 1.4. This arrangement leads to thinking of 6 as two rows of 3, or 6 = 3 + 3. Reflect 1.2 explores extending this thinking to other arrangements of dots.

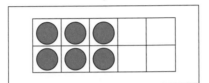

Fig. 1.4. Another way to show 6

Reflect 1.2

What number relationships might students perceive from the standard arrangements of dots on a number cube?

Example 2 lends itself to a different counting strategy:

Example 2: Sari has 5 apples. Three are red. The rest are yellow. How many of Sari's apples are yellow?

One way of using counting to solve this problem is to lay out 5 counters, separate (perhaps by circling) the 3 that represent red apples, and then count the remaining counters. Figure 1.5 depicts this situation.

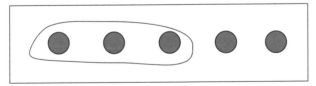

Fig. 1.5. Counters representing 5 with 3 as one part

We might also solve this problem by "counting on." We could lay out 3 counters for the 3 red apples and then count on until we had counters for 5 apples, as shown in figure 1.6.

Fig. 1.6. Counting on to represent 5 with 3 as one part

Alternatively, we might represent the problem by "counting back." We could start with 5 counters and then count back 3 for the 3 red apples, as illustrated in figure 1.7. Consider the question in Reflect 1.3 to compare "counting on" and "counting back."

Fig. 1.7. "Counting back" to represent 5 with 3 as one part

Reflect 1.3

Why is "counting back" so much more difficult than "counting on"?

The inverse relationship of addition and subtraction

Essential Understanding 1b. *Subtraction has an inverse relationship with addition.*

The chart in figure 1.8 shows the input and the output for the algebraic rule "add 2." The output number is always two more than the input number.

Rule: Add 2	
Input	Output
1	3
5	7
8	10
11	13

Fig. 1.8. Input/output table

The relationship in figure 1.8 is a *function* and can be represented by a function machine, as shown in figure 1.9. We can reverse the action of adding 2 by subtracting 2.

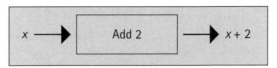

Fig. 1.9. A function machine for "add 2"

Example 2, about Sari's apples, showed a problem situation that some people would represent by addition, while others would use subtraction. The "counting on" subtraction strategy described above is grounded in the fact that $5 - 3 = \square$ is equivalent to $5 = 3 + \square$. The result of subtracting b from a, $a - b$, is formally defined as the number y where $a = b + y$. This definition builds logically on what students already know about addition, demonstrating why some problems can be solved by either operation.

Understanding the relationship between addition and subtraction reduces the number of facts that students must "know" by giving them a consistent, reliable strategy for subtraction: use the related addition fact. These related facts then form fact families. The third column in figure 1.10 shows a more formal algebraic description of a fact family.

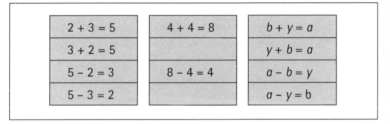

Fig. 1.10. Examples of fact families

The language describing subtraction is often very difficult for students to comprehend and use correctly. We may read the expression $5 - 3$ in many ways. Thinking about this expression in terms of parts and wholes may be helpful, since 5 (the *minuend*) is the whole, and 3 (the *subtrahend*) is a part. "Five minus 3" is the way that many adults would read the expression. Students might read it as "5 take away 3," but they might also say it as "3 taken away from 5." The same expression might be read either as "5 subtract 3" or as "3 subtracted from 5." It also might be read as "5 less 3" or as "3 less than 5." Note that in these phrases, the order of the numbers shifts, and some expressions include a preposition (*from* or *than*). It is very difficult for children to distinguish differences among the meanings of these phrases, and this confusion leads them to make frequent reversal errors. Chapter 3 describes strategies for helping students make sense of actions in word problems and the language of addition and subtraction.

For a discussion of the inverse relationship between multiplication and division, see *Developing Essential Understanding of Multiplication and Division for Teaching Mathematics in Grades 3–5* (Otto et al. 2011).

For a discussion of using appropriate terminology and representing word problems with multiplication and division, see *Developing Essential Understanding of Multiplication and Division for Teaching Mathematics in Grades 3–5* (Otto et al. 2011).

Representing situations with addition or subtraction

 Essential Understanding 1c. Many different problem situations can be represented by part-part-whole relationships and addition or subtraction.

Part-part-whole relationships show how two numbers—the parts— are related to a third number—the whole. For example, 2 and 3 are parts of the whole, 5. Addition and subtraction situations have been analyzed extensively and categorized in several different ways (see, for example, Carpenter [1985] and Schifter, Bastable, and Russell [2000]). The Common Core State Standards for Mathematics (Common Core State Standards Initiative 2010) include a table (adapted from the National Research Council [2009, pp. 32, 33]; see fig. 1.11) showing one way of categorizing different types of problems.

The first two types of problem situations identified in the table—"add to" and "take from"—involve actions. "Adding to" problems involve *increasing* by *joining*, while "taking from" problems involve *decreasing* or *separating*. Each of these situations can be further categorized by considering what information must be found (the *result* of the action, the *change*, or the *start*). Other problem situations do not change the amounts in any set; these "no action" situations may involve *putting together* collections of objects, *taking apart* a collection of objects, or *comparing* two collections of objects.

This section considers each of these types of situations by using part-part-whole representations. Reflect 1.4 invites further thinking about such situations.

Reflect 1.4

How is the language used to describe a situation related to its concrete representation?

Why is it important to understand what seem to be subtle distinctions in word problems?

Examples 1 and 2 have already illustrated two different types of problem situations, one of which was an action situation and the other of which was not.

- In example 1, Max began with 2 apples and then picked 5 more. The concrete representation shows a group of 2 counters and then a group of 5 counters. This "add to" action problem involves an initial quantity (*start*) and then an action that

	Result Unknown	Change Unknown	Start Unknown
Add to	Two bunnies sat on the grass. Three more bunnies hopped there. How many bunnies are on the grass now? $2 + 3 = ?$	Two bunnies were sitting on the grass. Some more bunnies hopped there. Then there were five bunnies. How many bunnies hopped over to the first two? $2 + ? = 5$	Some bunnies were sitting on the grass. Three more bunnies hopped there. Then there were five bunnies. How many bunnies were on the grass before? $? + 3 = 5$
Take from	Five apples were on the table. I ate two apples. How many apples are on the table now? $5 - 2 = ?$	Five apples were on the table. I ate some apples. Then there were three apples. How many apples did I eat? $5 - ? = 3$	Some apples were on the table. I ate two apples. Then there were three apples. How many apples were on the table before? $? - 2 = 3$

	Total Unknown	Addend Unknown	Both Addends Unknown
Put Together/ Take Apart	Three red apples and two green apples are on the table. How many apples are on the table? $3 + 2 = ?$	Five apples are on the table. Three are red and the rest are green. How many apples are green? $3 + ? = 5, 5 - 3 = ?$	Grandma has five flowers. How many can she put in her red vase and how many in her blue vase? $5 = 0 + 5, 5 = 5 + 0$ $5 = 1 + 4, 5 = 4 + 1$ $5 = 2 + 3, 5 = 3 + 2$

	Difference Unknown	Bigger Unknown	Smaller Unknown
Compare	("How many more?" version): Lucy has two apples. Julie has five apples. How many more apples does Julie have than Lucy?	(Version with "more"): Julie has three more apples than Lucy. Lucy has two apples. How many apples does Julie have?	(Version with "more"): Julie has three more apples than Lucy. Julie has five apples. How many apples does Lucy have?
	("How many fewer?" version): Lucy has two apples. Julie has five apples. How many fewer apples does Lucy have than Julie? $2 + ? = 5, 5 - 2 = ?$	(Version with "fewer"): Lucy has 3 fewer apples than Julie. Lucy has two apples. How many apples does Julie have? $2 + 3 = ?, 3 + 2 = ?$	(Version with "fewer"): Lucy has 3 fewer apples than Julie. Julie has five apples. How many apples does Lucy have? $5 - 3 = ?, ? + 3 = 5$

Fig. 1.11. Common addition and subtraction situations. Adapted from the Common Core State Standards for Mathematics (Common Core State Standards Initiative 2010, p. 88, from the National Research Council [2009, pp. 32, 33]).

joins to or *increases* that quantity, with the *result* unknown. It also indicates a part-part-whole relationship in which one part is the initial quantity and the other part is the quantity that is added (the *change*). The whole is the unknown *result* in this case. The common symbolic representation for this problem involves addition: $2 + 5 = \square$.

• In example 2, Sari has 5 apples, 3 of which are red. The concrete representation of this problem begins with 5 counters, which are then considered as two parts. This part-part-whole problem involves no action; it is static, with no apples added or taken away. The whole and one of the parts are given. The other part is what must be found. This *take-apart* problem might be considered to be a *missing addend* problem, because it can be represented as 5 = 3 + □. Note that writing the number sentence in this order models the situation directly and helps students understand the meaning of the equals sign as indicating that both sides represent the same amount. This missing addend problem situation can also be represented as subtraction: 5 − 3 = □.

The next five examples illustrate different types of "action" problems. Examples 3 and 4 involve "adding to" by *joining* or *increasing*, while examples 5–7 involve "taking from" by *separating* or *decreasing*. Sample situations of all types (that is, with the *result*, the *change*, or the *start* unknown) are provided.

Example 3 presents one type of "add to" situation:

Example 3: Juanita has 2 cookies. How many more cookies does she need to have 5 cookies?

Making a concrete representation of this problem involves beginning with 2 counters and then adding counters by a "counting on" strategy until there are 5 counters, as shown in figure 1.12. This "add to" problem involves an initial quantity (*start*) and then an action that *increases* that quantity; it entails *joining* an unknown number of cookies (the *change*) on to the starting set of 2 cookies. The initial quantity, or *start* (2), and the unknown *change* are the parts, and the *result* (5) is the whole. This problem is also a missing addend problem, which can be represented by either addition or subtraction: 2 + □ = 5 or 5 − 2 = □.

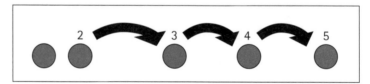

Fig. 1.12. Counters representing counting on from 2 to find a total of 5

Example 4 illustrates a different category of "add to" problem:

Example 4: Jose had some marbles. Angel gave him 3 marbles. Now he has 5 marbles. How many marbles did Jose have to start with?

Representing this "add to" problem concretely is difficult, since the initial quantity (*start*) is unknown. One way to model the problem concretely is to begin with several counters, perhaps in a container, and then make a separate group of 3 counters (to represent the marbles that Angel gives Jose), as shown in figure 1.13.

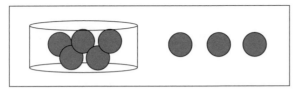

Fig. 1.13. Counters representing an unknown amount and 3 more

After Jose receives 3 marbles from Angel, he has 5 marbles in all. So, to continue to model the problem, we might add counters from the first group to the ones in the second group until we reach 5, as shown in figure 1.14.

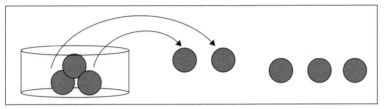

Fig. 1.14. Counters representing an unknown start and 3, with a whole of 5

Consider another concrete model, which requires working backward from the whole. This model involves setting out 5 counters (to represent Jose's total number of marbles) and then considering the parts, as suggested in figure 1.15 by the circling of 3 (to represent Jose's marbles from Angel) as one part of the whole. This "join" problem involves an unknown *start* (a part), an action *increasing* that quantity (a *change* of 3, or a part), and a *result* (the whole, 5). It is a missing addend problem, which can be represented either as addition or subtraction: $\Box + 3 = 5$, or $5 - 3 = \Box$.

Fig. 1.15. Counters representing 5, with 3 as one part

Example 5 turns the discussion to subtraction situations:

Example 5: Ben has 5 cookies. He eats 2 cookies. How many cookies does Ben have left?

This "take-away," or "separation," situation is familiar and easy to represent concretely. We might simply lay out 5 counters and then remove 2 of them. This action is frequently shown pictorially by crossing out counters, as illustrated in figure 1.16.

Fig. 1.16. Counters representing 5 take away 2

Note that in our model, we can remove any two counters, not just the ones on the right (or the left). Figure 1.17 shows an alternative way in which we could remove two counters.

Fig. 1.17. Another way to show 5 take away 2

This "take from" problem involves an initial quantity, or *start*, which is the whole. Then the problem describes a *decrease*, which *takes away*, or *separates*, a part, and the problem solver must find the missing part, or *result*. The situation depicted is a part-part-whole situation, with one part as the unknown. Notice that this subtraction story still depicts a part-part-whole relationship and still is in the form of *start–action* or *change–result*, but now the start is the whole, and the result is a part. This action situation is most often represented by subtraction: $5 - 2 = \square$.

Example 6 illustrates a different type of "take from" situation:

> **Example 6:** Mary has 5 flowers. She gives some to Anya. She has 2 flowers left. How many flowers did Mary give to Anya?

To model this "take from" problem concretely, we might start with 5 counters (see fig. 1.18a) and remove counters until only 2 are left (see fig. 1.18b). Like example 5, this problem involves a *start*, or initial quantity, which is the whole, but then it describes *decreasing* that quantity by *taking away* or *separating* an unknown part (the *change*) to leave 2 counters (the *result*, or remaining part). This action situation can be represented most directly from the context and modeled as $5 - \square = 2$. The addition number sentences $5 = 2 + \square$ and $5 = \square + 2$ are equivalent mathematically but are rarely used in this context.

Example 7 illustrates the third type of "take from" situation:

> **Example 7:** EJ has some books. He gives 2 books away. He has 3 left. How many books did EJ have to start with?

To represent this "take from" situation concretely, we might set out one group of 2 counters and another group of 3 counters and then combine these parts to find the whole. This interpretation of the

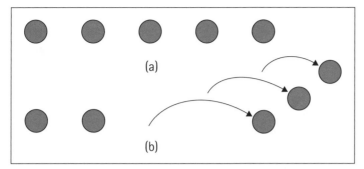

Fig. 1.18. Counters showing (a) a *start* of 5 and (b) a *result* of 2 left
after an *action* ("take away") separates some from 5

situation does not correspond to the way in which the situation is described, however, and thus it can be particularly difficult for students to understand. It requires working backward, putting together rather than taking away. To make a more basic concrete representation, we might start with a pile of counters and separate 2 from it (see fig. 1.19a). Then we might take counters away from the pile until only 3 are left (see fig. 1.19b).

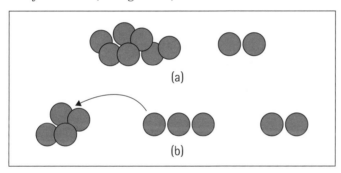

Fig. 1.19. Counters representing taking 2 from an unknown quantity,
with 3 remaining

This is a "take away," or "separate," action situation in which the *start*, or initial quantity, is unknown. The action involves *decreasing* the initial quantity by 2, leaving a *result* of 3. The situation may be represented symbolically as $\square - 2 = 3$, or as $2 + 3 = \square$. Note that the first number sentence is tied more closely to the context and model. Reflect 1.5 draws attention to how numbers can be used in different types of problems.

Reflect 1.5

Using 2, 7, and 9, write "add to" and "take from" problems with the *result* unknown, the *change* unknown, and the *start* unknown.

Six of the first seven examples have included an action ("add to" or "take from") and have illustrated how the unknown can be either a part or a whole in addition and subtraction stories. Not all addition and subtraction situations involve adding to or taking away, however, as example 2 demonstrated. The next five examples illustrate other no-action contexts for addition and subtraction. In these situations, the classification of the unknown as the *start*, *change*, or *result* is not appropriate. Deeply understanding addition and subtraction involves recognizing that we can use one or both of these operations to represent situations such as the following, while also realizing that an equation such as 7 – 3 = □ can represent very different situations. Students need to be able to model each type of situation directly, but they do not need to know the labels for each type of problem.

The situation in example 8 involves putting collections of objects together:

Example 8: Ming has 3 red balls and 2 blue balls. How many balls does he have in all?

A concrete representation of this problem would involve setting out two groups of counters (the parts), with the unknown whole. This "putting together" situation can be represented by the number sentence 3 + 2 = □. In this situation, the total is unknown.

Example 9 presents a situation that involves a comparison of collections of objects:

Example 9: Karen has 5 books. Jenny has 2 books. How many more books does Karen have than Jenny?

This "comparing" situation might be represented concretely, as shown in figure 1.20, by setting out 5 counters to show Karen's books and 2 counters to show Jenny's books, and then matching them up. In this case, the *difference* is unknown. The larger quantity (the number of Karen's books) can be considered as the whole in this problem, with the number of books that Jenny has as one part and the number of other books (the *difference*) as the other part. The situation can be represented symbolically by using either subtraction or addition: 5 – 2 = □ or 2 + □ = 5.

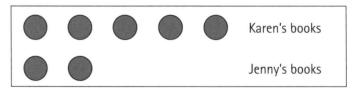

Fig. 1.20. Counters representing 5 compared with 2

The situation in example 10 also involves comparing:

Example 10: What is the length, in inches, of the line segment in figure 1.21?

In this situation, we can find the length of the line by determining the distance between 2 and 5 on the number line, or the number of units between those two numbers. We can consider the situation of finding this difference as a comparison situation, in which we know the whole (where the segment ends on the right), and we know one part (where the segment starts on the left), and we must find the part between the two (the *difference*). We can represent this situation symbolically by using either addition or subtraction: $2 + \square = 5$ or $5 - 2 = \square$.

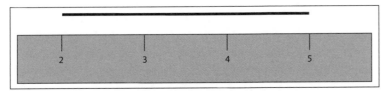

Fig. 1.21. Length as a comparison situation

The idea of using a number line to represent comparison situations is valuable. Reflect 1.6 explores possibilities.

Reflect 1.6

Think of some other situations involving comparisons.

How are these situations like finding the length of an object by using a ruler?

How are these situations like finding the difference between two numbers by using a number line?

Example 11 shows a different type of comparison situation:

Example 11: Jim has 2 marbles. Sarah has 3 more than Jim. How many marbles does Sarah have?

This example presents a comparison situation in which the difference is known and the larger quantity is unknown. Figure 1.22 shows that this situation can be represented concretely by laying out 2 counters in one row (the starting part) and then setting out counters to match the first 2 and 3 more (the other part).

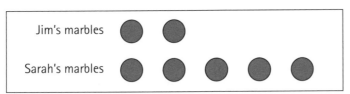

Fig. 1.22. Counters representing 3 more than 2

To answer the question, we must find the number of marbles that Sarah has (the whole). The number of marbles that Jim has composes one part, and the number of extra marbles that Sarah has composes the other part. The usual symbolic representation is the number sentence $2 + 3 = \square$, but the situation can also be represented by subtraction as $\square - 2 = 3$.

Example 12 shows yet another type of comparison situation:

Example 12: Ellen has 5 dolls. She has 2 more than Cindy. How many dolls does Cindy have?

Like the situation in example 11, this is also a comparison situation in which the difference is known, but this time the *smaller* quantity is unknown. One way to represent this situation concretely is shown in figure 1.23, which presents two rows of 5 counters, with the difference (2) removed from Cindy's set of dolls. If the larger number of dolls is considered as the whole, then the difference is one part, and the other part is the number of dolls Cindy has.

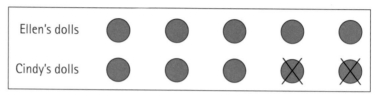

Fig. 1.23. Counters representing 2 less than 5

 Essential Understanding 1*d*

Part-part-whole relationships can be expressed by using number sentences like a + b = c or c − b = a, where a and b are the parts and c is the whole.

Note that an alternative way to model this situation concretely might simply be to use one set of 5 counters (Ellen's dolls) and then remove the difference. (Alternative representations will be discussed further in the next section, in connection with Essential Understanding 1*d*.) The symbolic representation might be either subtraction ($5 - 2 = \square$) or addition ($5 = \square + 2$).

Each of the twelve examples presented so far illustrates a slightly different problem situation, but each can be interpreted as a part-part-whole relationship. Each can be modeled both concretely and pictorially. Each can also be represented symbolically by using either addition or subtraction.

Part-part-whole relationships and number sentences

Essential Understanding 1*d*. *Part-part-whole relationships can be expressed by using number sentences like a + b = c or c − b = a, where a and b are the parts and c is the whole.*

In each of the examples discussed up to this point, we have seen that we can describe the problem situation as involving two parts that make up one whole. In each case, we have found that addition or subtraction (or both) is an appropriate operation to use to solve the problem. Each of these examples also meets two other important criteria:

1. The parts are *disjoint*; that is, they do not overlap with each other.

2. The parts are *exhaustive*; that is, the whole contains no objects other than the parts.

The next two examples show situations that may not satisfy these criteria.

Consider the seemingly simple situation in example 13:

Example 13: Two families are going to the shore. There are 5 people in one family and 6 people in the other family. How many people are going to the shore?

Most people encountering this problem would think that 11 people are going to the shore, as shown in figure 1.24a, but this may not be the correct answer. Suppose that one of the families includes a daughter of the other. Then, since the two groups overlap, as in figure 1.24b, we cannot just add to find the total. When adding or subtracting, the parts must be non-overlapping, or *disjoint*, sets. Each part is a set, and the union of these sets is the whole.

For more on these ideas about composing and decomposing quantities and part-whole relationships, see *Developing Essential Understanding of Number and Numeration for Teaching Mathematics in Prekindergarten–Grade 2* (Dougherty et al. 2010).

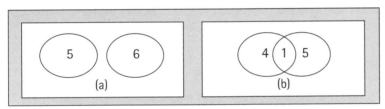

Fig. 1.24. Showing two families as (a) disjoint groups and (b) overlapping groups

Example 14 presents a situation involving possible arrangements or combinations of parts in the whole:

Example 14: Hal has 5 balls. Some are red, and some are blue. How many balls of each color might Hal have?

In this situation, Hal may have some yellow or green balls, or balls of other colors, in addition to the red and blue ones. However, if we assume that Hal has *only* red and blue balls, then we must find all of the possible ways to "make 5." Looking at the number 5 as the whole and determining all of the ways in which 5 might be split, or decomposed, into two parts, as shown below, lays the foundation for understanding addition and subtraction:

$$5 = 4 + 1 \qquad 5 = 3 + 2 \qquad 5 = 2 + 3 \qquad 5 = 1 + 4$$

Beyond knowing the different ways to decompose a particular number, such as 5, lies knowing what to expect when we decompose any number, as suggested in Reflect 1.7.

Reflect 1.7

What patterns do you see in the number sentences:

5 = 4 + 1,   5 = 3 + 2,   5 = 2 + 3,   and   5 = 1 + 4?

Why is it important to include both 5 = 4 + 1 and 5 = 1 + 4?

List all of the ways to make 10 as the sum of two counting numbers.

How many ways are there?

Representing situations in different ways

Essential Understanding 1e. *The context of a problem situation and its interpretation can lead to different representations.*

In most of the examples discussed previously, more than one number sentence is a reasonable choice for representing a problem situation symbolically. Often, the more appropriate choice depends on the way in which a problem situation is interpreted.

Consider the situation in example 15:

Example 15: Ana has 8 postcards to mail and only 6 postcard stamps. How many more stamps does she need?

One interpretation of this problem might be that Ana has 6 stamps and needs to have 8, so what number does she need to add to 6 to get 8? Under this interpretation, the situation represents a "putting together" part-part-whole relationship, where one part consists of the 6 stamps that Ana has, and the other part consists of the number of stamps that she needs. Ana's situation can thus be interpreted as a missing addend situation that leads to the number sentence 6 + □ = 8. Alternatively, this situation might be interpreted as a one-to-one correspondence, or a comparison situation: how many more than 6 is 8? In this case, the appropriate symbolic representation might be 8 – 6 = □. A child who does subtraction by thinking about a related addition fact might find the missing addend representation to be more useful than the equivalent subtraction one. Either addition or subtraction is an appropriate way to express a part-part-whole relationship.

In some of the number sentences discussed above, the whole appeared on the left side of the equals sign rather than the right side. It does not matter on which side of the equals sign we place the whole, since the expressions on each side are equivalent to each other. Try to suspend this understanding temporarily to take a child's perspective as you respond to the question in Reflect 1.8.

Reflect 1.8

Many children think that the equals sign always comes before the answer.

How might such a child complete the following number sentence:

2 + 6 = □ + 4?

Procedures for Adding and Subtracting: Big Idea 2

Big Idea 2. *The mathematical foundations for understanding computational procedures for addition and subtraction of whole numbers are the properties of addition and place value.*

Whether we compute by using traditional paper-and-pencil algorithms, mental math, estimation, or invented algorithms, the properties of addition, along with place value, provide the basis for our understanding of each procedure. Examination of the essential understandings related to Big Idea 2 highlights this fact.

The commutative and associative properties

Essential Understanding 2a. *The commutative and associative properties for addition of whole numbers allow computations to be performed flexibly.*

 Essential Understanding 1*d*

Part-part-whole relationships can be expressed by using number sentences like a + b = c or c − b = a, where a and b are the parts and c is the whole.

In example 14, discussed earlier in connection with Essential Understanding 1*d*, Hal has 5 balls, some of which are red and some of which are blue. Our list of possible combinations of red and blue balls shows that the expressions 1 + 4 and 4 + 1 can have different interpretations in this case, since one number represents the number of red balls, and the other, the number of blue balls. More generally, however, it does not matter whether we put the red balls or the blue ones first, since 1 + 4 gives the same result as 4 + 1. The commutative property for addition of whole numbers is stated generally as $a + b = b + a$. Although it may not be necessary for students to state the commutative property by using variables or even to recognize it by that name, it is essential that they be able to explain and use the property in solving problems, as illustrated in the situation with the balls.

The commutative property simplifies addition greatly, especially as children are learning their number facts. Because 3 + 8 is the same as 8 + 3, they can find the sum by using the "counting on" strategy, starting with the larger number. Use of the commutative property cuts the number of addition facts that children must memorize from 100 to 55. The addition table in figure 1.25 illustrates this fact; it is symmetric about the diagonal (shaded), and the answers (or wholes) for $a + b$ and $b + a$ appear directly across the diagonal from each other.

Understanding and being able to use the commutative property for addition are also very important in doing mental math with larger numbers. For example, to find 17 + 135 mentally, starting

	0	1	2	3	4	5	6	7	8	9
0	0	1	2	3	4	5	6	7	8	9
1	1	2	3	4	5	6	7	8	9	10
2	2	3	4	5	6	7	8	9	10	11
3	3	4	5	6	7	8	9	10	11	12
4	4	5	6	7	8	9	10	11	12	13
5	5	6	7	8	9	10	11	12	13	14
6	6	7	8	9	10	11	12	13	14	15
7	7	8	9	10	11	12	13	14	15	16
8	8	9	10	11	12	13	14	15	16	17
9	9	10	11	12	13	14	15	16	17	18

Fig. 1.25. Addition facts table

with 135 and then counting on 17 is much easier than starting with 17 and then counting on 135.

Example 16 illustrates the usefulness of another property of addition:

Example 16: Mrs. Brown rolls three number cubes each day for her students to add. Today she rolled the numbers 5, 6, and 4. What is the sum?

One way to find the sum is to add the numbers in order (see fig. 1.26). Another way is to add the last two numbers first to make 10 (see fig. 1.27). The associative property for addition allows us to re-group addends; it is generally stated formally as follows:

$$a + (b + c) = (a + b) + c$$

Being able to use the associative property fluently is important in developing good number sense.

In developing an understanding of the counting numbers, students learn that a number can be decomposed as a sum of two parts in many ways; for example, $8 = 5 + 3$, $8 = 4 + 4$, and $8 = 2 + 6$. Combining this notion of decomposing a number with the commutative and associative properties is foundational to most addition and subtraction fact strategies. One strategy that children use for

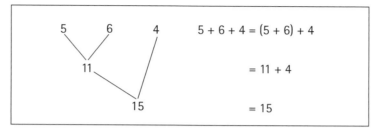

Fig. 1.26. Adding (5 + 6) + 4

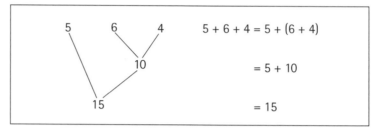

Fig. 1.27. Adding 5 + (6 + 4)

the addition of small numbers is the "make 10" strategy. To model this strategy concretely, consider the numbers 8 and 5, as shown on ten-frames in figure 1.28.

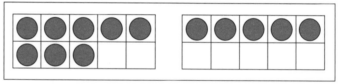

Fig. 1.28. Modeling counting on a ten-frame

To find the sum of 8 and 5, we can split 5 into 2 + 3. We can move 2 of the counters on the right to the ten-frame on the left. Then, as shown in figure 1.29, we can see that 8 + 2 is 10, and 10 + 3 = 13. Representing our model formally in writing, we can see the decomposing of 5, followed by the use of the associative property for addition:

$$8 + 5 = 8 + (2 + 3) = (8 + 2) + 3 = 10 + 3 = 13.$$

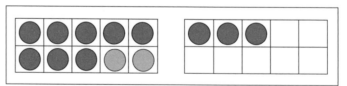

Fig. 1.29. Using the "make 10" strategy

For a discussion of the commutative and associative properties of multiplication and division and their roles in computation strategies, see *Developing Essential Understanding of Multiplication and Division for Teaching Mathematics in Grades 3–5* (Otto et al. 2011).

We can use a similar strategy to simplify computations with larger numbers. For example, one way of using the commutative and associative properties of addition to find 78 + 26 involves working with 100 as a benchmark number by grouping 75 and 25:

$$78 + 26 = (3 + 75) + (25 + 1) \quad \text{Decomposition of each addend}$$
$$= 3 + (75 + (25 + 1)) \quad \text{Associative property for addition}$$
$$= 3 + ((75 + 25) + 1) \quad \text{Associative property for addition}$$
$$= 3 + (100 + 1) \quad \text{Addition}$$
$$= 3 + (1 + 100) \quad \text{Commutative property for addition}$$
$$= (3 + 1) + 100 \quad \text{Associative property for addition}$$
$$= 4 + 100 = 104 \quad \text{Addition}$$

Alternatively, we might use 80 as a benchmark number:

$$78 + 26 = 78 + (2 + 24) \quad \text{Decomposition of the second addend}$$
$$= (78 + 2) + 24 \quad \text{Associative property for addition}$$
$$= 80 + 24 \quad \text{Addition}$$
$$= 80 + (20 + 4) \quad \text{Decomposition of the second addend}$$
$$= (80 + 20) + 4 \quad \text{Associative property for addition}$$
$$= 100 + 4 = 104 \quad \text{Addition}$$

In fact, we can find many ways to obtain the sum 78 + 26 by working with multiples of 10 or 100 as benchmark numbers and using the properties of addition. Reflect 1.9 explores the use of this strategy.

Reflect 1.9

Find at least two ways in which the commutative and associative properties of addition can help you find each of the following sums:

$$97 + 105 \qquad 347 + 454$$

The associative property is also the basis for the strategy of using known facts to find sums of one-digit numbers. This strategy is, in fact, a more general version of the "make 10" strategy but can also be used with other facts. For example, suppose that we need to find 7 + 8, and we know that 7 + 4 = 11. Then we can break 8 into 4 + 4 and add each part separately:

$$7 + 8 = 7 + (4 + 4) = (7 + 4) + 4 = 11 + 4 = 15$$

Another type of problem that students frequently solve by using this strategy involves "near doubles"—two addends that nearly duplicate each other. Because students usually remember addition facts for doubles, such as 7 + 7 = 14, they can use these facts to help find other sums. For example,

$$7 + 8 = 7 + (7 + 1) = (7 + 7) + 1 = 14 + 1 = 15.$$

The associative property for addition can also be extremely helpful to students in grade 3 or 4 when they are adding a column of several numbers. Consider the following sum:

$$
\begin{array}{r}
44 \\
65 \\
36 \\
77 \\
28 \\
+82 \\
\end{array}
$$

Students can use the "make 10" strategy to look for number combinations in the ones column that add to 10:

$$4 + 6 = 10 \quad \text{and} \quad 8 + 2 = 10$$

Note that this approach to the digits in the ones column requires the use of both the associative and commutative properties:

$$
\begin{aligned}
(4 + 5 + 6) + 7 + (8 + 2) \ &= (4 + (6 + 5)) + 7 + 10 \\
&= ((4 + 6) + 5) + 7 + 10 \\
&= 10 + (5 + 7) + 10 \\
&= 10 + (12 + 10) \\
&= 10 + (10 + 12) = (10 + 10) + 12 = \\
&\quad 20 + 12 = 32
\end{aligned}
$$

Continuing to the tens column in the same problem, students can use an analogous "make 100" process, first looking for combinations that make 100: $40 + 60 = 100$, $30 + 70 = 100$, and $20 + 80 = 100$. They will now have $300 + 32 = 332$. Note that the "make 100" process illustrated here has many variations, all of which are equally correct and valid.

Subtraction: Neither commutative nor associative

Essential Understanding 2b. Subtraction is not commutative or associative for whole numbers.

Subtraction does not have the same properties as addition. Consider the two expressions $5 - 2$ and $2 - 5$. Although both involve finding the distance between 2 and 5, they do not have the same meaning. We might read the first as "5 minus 2" or "the difference between 5 and 2," and the second as "2 minus 5" or "the difference between 2 and 5." In the first case, we might interpret the number sentence to mean, "Start with 5 counters, and then take away 2," leaving 3 counters. And in the second case, we might interpret the sentence

2 – 5 in the same fashion to mean, "Start with 2 counters and then take away 5." Even young children can understand that in this situation, they can take away 2 but then still need to take away 3 more; they learn much later that doing so is possible, but only by expanding to negative numbers, when they discover that the correct answer is –3. Subtraction does *not* have a commutative property; the order of the terms in a subtraction expression is important. Reflect 1.10 investigates a closely related false claim that students frequently make about subtraction.

Reflect 1.10

What misunderstandings might students have if they make the assertion, "You can't take a bigger number from a smaller one?"

Subtraction also does not have an associative property. Because a property must hold for every possible combination of numbers, we can show that subtraction is not associative by producing a single example that does not work. Consider the following two ways in which we might use subtraction with the numbers 15, 8, and 5:

$$(15 - 8) - 5 = 7 - 5 = 2$$
$$15 - (8 - 5) = 15 - 3 = 12$$

The values of the two expressions $(15 - 8) - 5$ and $15 - (8 - 5)$ are not the same, so subtraction is not associative.

The "make 10" strategy, shown earlier as an application of the associative property of addition of whole numbers, can be applied to subtraction. To find 15 – 7, we can start with 15, shown with counters on two ten-frames as in figure 1.30a. By thinking of 7 as 5 + 2, we first take 5 away and have 10 left (fig. 1.30b). Then we take away 2 more counters (fig. 1.30c): 10 – 2 = 8.

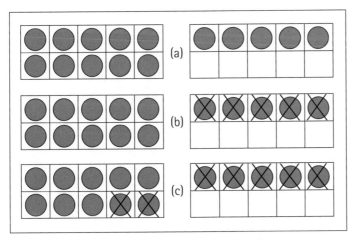

Fig. 1.30. Using the "make 10" strategy for 15 – 7

To show formally and symbolically why the "make 10" strategy works with subtraction requires that we use integers and their properties, which are familiar to us, if not to our students. Note that such a justification is not appropriate for children in the primary grades, since it requires the use of negative numbers. In this justification, we subtract a quantity by adding the inverse, or opposite, of the number; we will call this procedure the "definition of subtraction." The opposite of a number is also equal to the product of the number and –1: $-x = -1(x)$. This property is called "multiplication by –1" in the analysis below.

$$c - (a + b) = c + -(a + b) \qquad \text{Definition of subtraction}$$
$$= c + (-1)(a + b) \qquad \text{Multiplication by –1}$$
$$= c + ((-1) \cdot a + (-1) \cdot b) \qquad \text{Distributive property for multiplication over addition}$$
$$= (c + (-1 \cdot a)) + (-1 \cdot b) \qquad \text{Associative property for addition}$$
$$= (c - a) - b \qquad \text{Multiplication by –1}$$

Thus, $15 - 7 = 15 - (5 + 2) = (15 - 5) - 2 = 10 - 2 = 8$.

As in the case of addition, we can extend the "make 10" strategy to a "make 100" or "make 1000" or "make a multiple of 10" strategy to subtract multi-digit numbers. Consider an example:

$$1002 - 17 = 1002 - (2 + 15) = (1002 - 2) - 15 = 1000 - 15 = 985$$

Solving mentally involves decomposing 17 as $2 + 15$ and then distributing the multiplication by –1.

Place value and composition

Essential Understanding 2c. *Place-value concepts provide a convenient way to compose and decompose numbers to facilitate addition and subtraction computations.*

For more on ideas about place value, see *Developing Essential Understanding of Number and Numeration for Teaching Mathematics in Prekindergarten– Grade 2* (Dougherty et al. 2010).

Composing and decomposing numbers involves combining and separating them to make parts and wholes. In our base-ten number system, when we accumulate 10 units, we can regroup them into a larger unit. Ten ones make 1 ten, and 10 tens make 1 hundred, and so on. The size of each new group determines the new place value. A digit thus represents a different quantity, depending on its location in a number. The digit 2 represents 2 ones in 562, 2 tens in 526, and 2 hundreds in 256.

Place value can be very helpful when we are using mental math to add and subtract numbers. For example, if we need to find

46 + 30, it is helpful to recognize that we are simply adding 3 tens to 46. We might then count on by 10s from 46: 56, 66, 76.

Similarly, to find 46 + 32, we might decompose each number into tens and ones. When doing mental math, many people work from left to right. We might think as follows:

46 + 32 = (40 + 6) + (30 + 2) = (40 + 30) + (6 + 2) = 70 + 8 = 78

Estimation relies heavily on place-value concepts. To estimate a sum or difference, we want to replace the given numbers with numbers that are close to them but allow us to do the computation mentally. Thus, to find 479 – 186, we might round each number to the nearest hundred: 500 – 200 = 300. We might round the numbers to the nearest ten, but 480 – 190 is difficult to compute mentally, so we might think of this as

480 – 200 + 10 = 280 + 10 = 290.

We might also recognize that 480 – 190 is very close to either 490 – 190 or 480 – 180. In each case, our understanding of place value is critical in thinking about the numbers flexibly and decomposing and composing them to simplify computation. Reflect 1.11 invites you to consider possibilities.

Reflect 1.11

Find these sums and differences by composing and decomposing the numbers in different ways:

48 + 23	93 – 38
376 + 127	738 – 129

Properties and algorithms

Essential Understanding 2d. Properties of addition are central in justifying the correctness of computational algorithms.

Procedures for finding sums and differences of multi-digit numbers are all based on the properties of addition, whether they involve mental math, estimation, standard paper-and-pencil algorithms, or other algorithms. The discussion of Essential Understanding 2d that follows first considers addition and subtraction on the number line and hundreds chart, and then it examines various addition and subtraction algorithms.

Working with a number line

Two useful models for developing understanding of addition and subtraction are a number line and a hundreds chart. In each case,

For a discussion of using the properties of multiplication and division to justify the correctness of algorithms, see *Developing Essential Understanding of Multiplication and Division for Teaching Mathematics in Grades 3–5* (Otto et al. 2011).

students extend their use of the "counting on" and "counting back" strategies to add and subtract two-digit numbers. Although a number line is not necessary for such reasoning, it can be helpful in visualizing the actions of addition or subtraction. For example, as shown in figure 1.31, to add 38 + 16 on the number line, a student might first add 38 + 10, then add 2 to get to a "nice" number (a multiple of 10), and finally add 4.

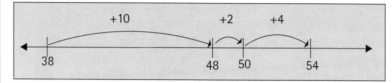

Fig. 1.31. A number line representing 38 + 16 as 38 + 10 + 2 + 4

Another way for a student to find this sum might be to begin at 38, add 2 to get to a "nice" number, and finish by adding 14, as illustrated in figure 1.32. In each case, the student is using decomposition and the associative property for addition as he or she takes the numbers apart and then puts them back together in different ways.

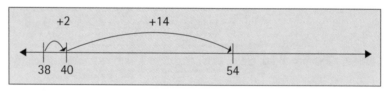

Fig. 1.32. Another number line representation for 38 + 16 as 38 + 2 + 14

We can subtract either by counting back or counting on. To find 62 – 25, one person might start at 62 and count back to subtract 20, and then count back to subtract 2 to get to a multiple of 10, and finally count back to subtract 3. An illustration of this strategy appears in figure 1.33.

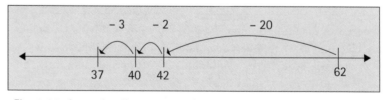

Fig. 1.33. A number line representing counting back to find 62 – 25

Another person might first mark both numbers on the number line and then start at 25 and count on, as illustrated in figure 1.34. In using this strategy, it is critical to keep track of the differences between the intermediate numbers. In the case shown in the figure, the difference is 5 + 30 + 2 = 37.

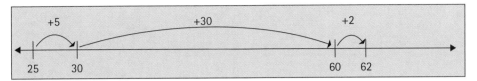

Fig. 1.34. A number line representation of counting on to find 62 – 25

Working with a hundreds chart

A hundreds chart (or 0–99 chart) like that shown in figure 1.35 shows place-value relationships by lining up the digits in the ones column. It is different from the number line in that it shows only the whole numbers, with no spaces between for fractions or decimals. By using the hundreds chart to add and subtract, students can build a visual model that can be very helpful in doing mental math.

1	2	3	4	5	6	7	8	9	10
11	12	13	14	15	16	17	18	19	20
21	22	23	24	25	26	27	28	29	30
31	32	33	34	35	36	37	38	39	40
41	42	43	44	45	46	47	48	49	50
51	52	53	54	55	56	57	58	59	60
61	62	63	64	65	66	67	68	69	70
71	72	73	74	75	76	77	78	79	80
81	82	83	84	85	86	87	88	89	90
91	92	93	94	95	96	97	98	99	100

Fig. 1.35. A hundreds chart

For example, to find 38 + 16, we can use the "counting on" strategy, beginning with the larger number, 38. As shown in figure 1.36, we might first add 10 by moving down one row, and then count on 2 to get to 50, and then count on 4 to finish up.

To find 62 – 25, we can either count back or count on. To count back, we start at 62 and count back 25, as shown in figure 1.37. To do so, we can count back 2 rows to 42 and then count back 5 ones to 37.

31	32	33	34	35	36	37	38	39	40
41	42	43	44	45	46	47	48	49	50
51	52	53	54	55	56	57	58	59	60

Fig. 1.36. A hundreds chart representing 38 + 16

31	32	33	34	35	36	37	38	39	40
41	42	43	44	45	46	47	48	49	50
51	52	53	54	55	56	57	58	59	60
61	62	63	64	65	66	67	68	69	70

Fig. 1.37. A hundreds chart representing 62 – 25 by counting back

To count on, we might start at 25, move down 3 rows to 55, and then count on 7 more to get to 62, as illustrated in figure 1.38. The difference is 30 + 7, or 37.

21	22	23	24	25	26	27	28	29	30
31	32	33	34	35	36	37	38	39	40
41	42	43	44	45	46	47	48	49	50
51	52	53	54	55	56	57	58	59	60
61	62	63	64	65	66	67	68	69	70

Fig. 1.38. A hundreds chart showing 62 – 25 by counting on

Alternatively, we might start at 25, move down 4 rows to 65 and then count back 3 to get to 62. The difference in this case is 40 – 3, or 37, and we are actually doing the following computation: $25 + 40 - 3 = 65 - 3 = 62$. Reflect 1.12 invites an analysis of this work with hundreds charts and number lines.

Reflect 1.12

What role do the properties of addition play in the computations performed on a number line and a hundreds chart?

The standard algorithm for paper-and-pencil addition

Place-value blocks provide a concrete way to represent the addition process known as "regrouping." Again consider the case of 38 + 16, as illustrated in figure 1.39. First, we lay out blocks to show 38 as 3 tens and 8 ones, and 16 as 1 ten and 6 ones. Next, since we have more than 9 ones, we add these together to make 14, or 1 ten (shown by the circled ones in the figure) and 4 ones. We record a 4 in the ones column of our addition, as shown on the right in the figure. Then we record the 1 ten as a 1 above the tens place in the number 38 in our computation, and we add the tens: 1 + 3 + 1 = 5. We record a 5 in the tens column, for a sum of 54.

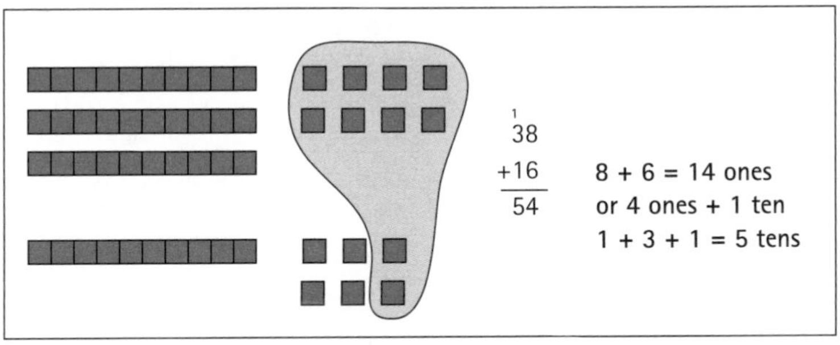
1
38
+16
54
8 + 6 = 14 ones
or 4 ones + 1 ten
1 + 3 + 1 = 5 tens

Fig. 1.39. Place-value blocks modeling the standard algoritm for addition

This process can be formally justified by using the commutative and associative properties:

38 + 16 = (30 + 8) + (10 + 6)	Place value
= 30 + (8 + 10) + 6	Associative property for addition (twice)
= 30 + (10 + 8) + 6	Commutative property for addition
= (30 + 10) + (8 + 6)	Associative property for addition (twice)
= 40 + 14	Addition
= 40 + (10 + 4)	Place value
= (40 + 10) + 4	Associative property for addition
= 50 + 4	Addition
= 54	Place value

Note that this formal justification is intended for teachers; students are unlikely to use this language or degree of formality in explaining their work.

Other algorithms for addition

Another method for adding multi-digit numbers uses very much the same properties as those shown formally above but does not involve regrouping at all. This method, the "partial sums" algorithm, adds each place value separately and then adds the sums. This procedure may begin either on the left, as in examples (a) and (b) in figure 1.40, or on the right, as in example (c).

```
                                         2385
                      128                1239
         38           258               +4528
        +16          +691               ─────
        ────         ────                  22
         40           900                 130
        +14           160                1000
        ────         + 17               +7000
         54          ─────              ─────
                     1017                8152


        (a)          (b)                 (c)
```

Fig. 1.40. Adding by using the partial sums algorithm

Yet another addition algorithm that students sometimes learn is called the "opposite change" (or "give and take" or "add-subtract") algorithm. In this procedure, the goal is to change one of the numbers to a multiple of 10 (or 100 or 1000) so that it is easy to add. We can accomplish this by adding a number to one of the addends and subtracting it from the other. This is similar to the "make 10" strategy discussed earlier, as demonstrated in the following simple example:

$$9 + 7 = (9 + 1) + (7 - 1) = 10 + 6 = 16$$

Figure 1.41 shows a slightly more advanced example; 4 is subtracted from the first number and added to the second to make a multiple of 10. A still more complex example in figure 1.42 uses two modifications instead of just one to determine the sum.

```
         38    -4     34
               →
        +16    +4    +20
        ────   →     ────
                      54
```

Fig. 1.41. Subtracting and adding 4 to simplify addition

$$428 \xrightarrow{+2} \quad 430 \xrightarrow{+70} \quad 500$$
$$+576 \xrightarrow{-2} \quad +574 \xrightarrow{-70} \quad +504$$
$$\overline{} \qquad \overline{} \qquad \overline{1004}$$

Fig. 1.42. Using two steps in the "opposite change" algorithm

Why does this procedure give a correct answer? Consider the general case of $a + b$, where we add and subtract a number c:

$(a + c) + (b - c) = a + (c + (b - c))$	Associative property for addition
$= a + ((c + b) + (-c))$	Associative property for addition and definition of subtraction
$= a + ((b + c) + (-c))$	Commutative property for addition
$= a + (b + (c + (-c)))$	Associative property for addition
$= a + (b + 0)$	Additive inverse for addition
$= a + b$	Addition identity

Two properties of addition are used here. The additive inverse property for addition states that any number x has an opposite, $-x$, such that the sum of the two numbers $x + (-x) = x - x = 0$. Later, students will paraphrase this property by saying, "The sum of any number and its opposite is zero." In the primary grades, however, since students do not use negative numbers, the idea is usually stated simply as, "Any number minus itself is zero." The addition identity is zero, the number that can be added to any other number without changing its value. We usually write this property of additive identity symbolically as $0 + x = x = x + 0$.

Students first use these properties as they learn their addition and subtraction facts, and then they continue to use them in the later grades when they solve equations and simplify algebraic expressions. Reflect 1.13 explores the use of the partial sums and opposite change algorithms.

For a discussion of using properties of multiplication and division to justify alternative strategies, see *Developing Essential Understanding of Multiplication and Division for Teaching Mathematics in Grades 3–5* (Otto et al. 2011).

Reflect 1.13

Use the "partial sums" and "opposite change" algorithms to find the following sums:

$$36 + 58 \qquad 296 + 375 \qquad 1357 + 8642$$

The standard algorithm for subtraction

Place-value blocks may be useful in illustrating the algorithm for subtraction that is most widely used at the present time. In using this procedure, we begin computing with the ones column and move to the left, regrouping as needed. An example appears in figure 1.43.

Representation with Place-Value Blocks	Symbolic Representation	Description in Words
	$\begin{array}{r}4\ 2\\ -1\ 7\\\hline\end{array}$	Represent 42 with place-value blocks.
	$\begin{array}{r}{}^{3}\cancel{4}{}^{1}2\\ -1\ 7\\\hline\end{array}$	Because we need to take away 7 ones but only have 2 ones, we regroup 1 ten as 10 ones, so now we have 3 tens and 12 ones.
	$\begin{array}{r}{}^{3}\cancel{4}{}^{1}2\\ -1\ 7\\\hline 5\end{array}$	Subtract the ones.
	$\begin{array}{r}{}^{3}\cancel{4}{}^{1}2\\ -1\ 7\\\hline 2\ 5\end{array}$	Subtract the tens.

Fig. 1.43. Place-value blocks modeling the standard algorithm for subtraction

This procedure, like the standard algorithm for addition, can be justified by using the properties of addition and place value:

$42 - 17 = (30 + 12) - (7 + 10)$	Facts and place value
$= (30 + 12) - 7 - 10$	$c - (a + b) = c - a - b$ (see discussion of Essential Understanding 2c)
$= (30 + 12) + (-7) + (-10)$	Definition of subtraction
$= 30 + (12 + (-7)) + (-10)$	Associative property for addition
$= 30 + 5 + (-10)$	Addition
$= 30 + (-10) + 5$	Commutative property for addition
$= 30 - 10 + 5 = 25$	Addition, subtraction, and definition of subtraction

With larger numbers, the standard algorithm usually alternates between regrouping each place value and subtracting it. A variation of this algorithm performs all regrouping first, beginning with the ones column, as illustrated in figure 1.44a, and then subtracts each place value, as illustrated in figure 1.44b.

Essential Understanding 2c
Place-value concepts provide a convenient way to compose and decompose numbers to facilitate addition and subtraction computations.

(a)				(b)			
6	11	10	11	6	11	10	11
~~7~~	~~2~~	~~1~~	~~1~~	~~7~~	~~2~~	~~1~~	~~1~~
-3	6	5	4	-3	6	5	4
				3	5	5	7

Fig. 1.44. Varying the standard algorithm by first (a) regrouping all place values, and then (b) subtracting

Other algorithms for subtraction

Many other algorithms exist for subtraction. Some are used in other cultures today, some were widely taught in the United States, Canada, or Mexico in the past, and others are the invention of students, past or present. Several of these alternative algorithms for subtraction are useful to highlight.

The first of these algorithms, often known as the "partial differences" algorithm, has two variations. In its basic form, this algorithm proceeds from left to right, with users subtracting each place value one by one. This procedure requires very good mental math skills, especially in making multiples of 10, 100, and 1000. Figure 1.45 shows the procedure for the problem 438 – 172. Each step appears in the center, with two different ways of recording the steps on each side. Some very able students (and adults) do these computations mentally, recording only the result.

438		438
-172		-100
338	438 - 100 = 338	338
-72	338 - 70 = 268	-70
268	268 - 2 = 266	268
-2		-2
266		266

Fig. 1.45. Using the partial differences algorithm to subtract 438 – 172

The second variation on the partial differences algorithm sometimes uses negative numbers. Users of this procedure compute each place value one at a time, beginning on the left. Figure 1.46 illustrates their process. If the top digit is larger than the bottom one (as in the ones column in the figure), then they add the difference onto the partial difference. If the top digit is smaller than the bottom one (as in the tens column), then they subtract the difference from the partial difference.

```
  438
 -172
 ————        400 – 100 = 300
  300
              Have 30, need to subtract 70,
  -40
 ————        so subtract 40 from previous difference: 300 – 40 = 260;
  260
              8 – 2 = 6, so need to add 6.
   +6
 ————
  266
```

Fig. 1.46. A variation on the partial differences algorithm

It may be easier to see how this algorithm works by looking at it horizontally:

$$438 - 172 = (400 + 30 + 8) - (100 + 70 + 2)$$
$$= (400 - 100) + (30 - 70) + (8 - 2)$$
$$= 300 - 40 + 6$$
$$= 266$$

The second step involves using the definition of subtraction and the associative and commutative properties for addition as well as the distributive property. Because this algorithm begins to lay a foundation for negative numbers and provides ways to subtract a larger digit from a smaller one, it is attractive to middle grades teachers. However, although students sometimes invent the algorithm on their own, it is rarely taught as the preferred algorithm for all students.

The next algorithm builds on the procedure for using a number line or hundreds chart for subtracting. This "counting on" procedure uses "nice" numbers to count up from the subtrahend (the part, or the number, being subtracted) to the minuend (the whole). This algorithm is also often called the "cashier's algorithm," since it offers an efficient way of making change. The following subtraction problem offers an example:

4000 – 2356	Start at 2356 and add 4 to get 2360	4
	Add 40 to get 2400	40
	Add 600 to get 3000	600
	Add 1000 to get 4000	1000
	Add the partial differences	1644

Note that this procedure does not require any regrouping. Reflect 1.14 explores the two alternative subtraction algorithms discussed so far.

Reflect 1.14

Use the "partial differences" and "counting on" algorithms to find each of the following differences:

$$73 - 38 \qquad 642 - 135$$
$$702 - 187 \qquad 2305 - 1268$$

Another algorithm, often called the "Austrian method," is dominant in some cultures and was widely used in the United States prior to 1940 (Ross and Pratt-Cotter 1999). This algorithm, also called the "additions," "equal addends," or "equal addition" method, involves adding the same number (10, 100, 1000, etc.) to both the minuend and the subtrahend, as illustrated in figure 1.47.

13	11	5 is larger than 1, so add 10 ones to 1 and 1 ten to 7 11 − 5 = 6
4 $\not{3}$ $\not{1}$		8 is larger than 3, so add 10 tens to 3 and 1 hundred to 2 13 − 8 = 5
$\not{2}^{3}$ $\not{7}^{8}$ 5		4 hundreds − 3 hundreds = 1 hundred
1 5 6		

Fig. 1.47. Using the Austrian method to subtract 438 − 275

An alternative notation for this algorithm sometimes shows these additions by placing 1s on diagonals between the numbers. Figure 1.48 illustrates this process for the subtraction 431 − 174. First a 1 is placed on the diagonal between the 1 in the ones column and the 7 in the tens column. This 1 represents the 1 ten to be added to both numbers. A 1 is also placed on the diagonal between the 3 in the tens column and the 1 in the hundreds column. This 1 represents the 100 to be added to both of these numbers. Subtraction proceeds as shown in the figure.

4 $_1$3 $_1$1	11 ones − 4 ones = 7 ones
−1 7 4	13 tens − 1 ten − 7 tens = 5 tens
2 5 7	4 hundreds − 1 hundred − 1 hundred = 2 hundreds

Fig. 1.48. The Austrian method for subtracting 431 − 174

Reflect 1.15 invites you to try out the Austrian method for subtraction. This algorithm is very similar to the next one—the last one discussed here—which involves making the same change to both the minuend and the subtrahend.

Reflect 1.15

Use the Austrian algorithm (also known as the "additions," "equal addends," or "equal addition" algorithm) to find each of the following differences:

$$73 - 38 \qquad 642 - 135$$

$$702 - 187 \qquad 2305 - 1268$$

The "same change" algorithm for subtraction is more general than the Austrian method, since any number can be added to the numbers being subtracted. Usually, numbers are chosen to simplify the problem and eliminate the need for regrouping. In the example in figure 1.49, a 6 is first added to get a 0 in the ones column in the number being subtracted. Then a 20 is added to get a multiple of 100. Now the subtraction is very easy. This method requires only addition and subtraction facts with sums to 10. Reflect 1.16 explores the use of this algorithm.

$$
\begin{array}{r}
431 \\
-174 \\
\hline
\end{array}
\;\overset{+6}{\rightarrow}\;
\begin{array}{r}
437 \\
-180 \\
\hline
\end{array}
\;\overset{+20}{\rightarrow}\;
\begin{array}{r}
457 \\
-200 \\
\hline
257
\end{array}
$$

Fig. 1.49. The "same change" algorithm for subtracting 431–174

Reflect 1.16

Use the "same change" algorithm to find each of the following differences:

$$73 - 38 \qquad 642 - 135$$

$$702 - 187 \qquad 2305 - 1268$$

One way to justify this procedure involves considering the difference of two numbers ($b - a$) as the distance between those points on a number line. If the same number — say, c — is added to both numbers, then both slide by the same distance on the number line, so they stay the same distance apart. Figure 1.50 illustrates this fact.

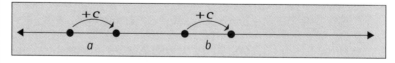

Fig. 1.50. Using the number line to represent the "same change" algorithm for subtraction

The properties of addition and place value can be used to show more formally why this procedure works. Suppose that in the case of two whole numbers a and b, with b being subtracted from a, a whole number c is added to both numbers.

$(a + c) - (b + c)$	$= (a + c) + -(b + c)$	Definition of subtraction
	$= (a + c) + (-1)(b + c)$	Multiplication by -1
	$= (a + c) + (-1)(b) + (-1)(c)$	Distributive property for multiplication over addition
	$= (a + c) + (-b) + (-c)$	Multiplication by -1
	$= a + (c + (-b)) + (-c)$	Associative property for addition
	$= a + (-b + c) + (-c)$	Commutative property for addition
	$= (a + (-b)) + (c + (-c))$	Associative property for addition
	$= (a - b) + (c - c)$	Definition of subtraction
	$= (a - b) + 0$	Additive inverse
	$= a - b$	Zero as additive identity

Thus, the "same change" algorithm is the result of generalizing an arithmetic pattern; adding the same number to both the minuend and subtrahend does not change the difference between the two numbers. Reflect 1.17 invites further investigation of the algorithm.

Reflect 1.17

Can you subtract the same number from both the minuend and subtrahend to simplify subtraction?

If so, why does this work?

Solve the following problems by adding the same number and then try to solve them by subtracting the same number.

$$73 - 59 \qquad 523 - 378$$
$$4321 - 1987$$

Consider your results, and if possible, discuss them with your colleagues.

Conclusion

To understand and use addition and subtraction fluently, students need to understand when to use each operation, how to write number sentences to match specific problem situations, and how to compute sums and differences. Students need to be able not only to do computations but also to explain why these procedures work, using words, diagrams, or models. The essential understandings described in this chapter constitute a solid foundation for teachers, enabling them to recognize addition and subtraction in its many contexts, to apply and validate a variety of strategies to solve addition and subtraction problems, and to use place value and strategies such as "make 10" and using known facts to help students make sense of basic facts and mental computations. For students in prekindergarten through grade 2, these ideas are essential in and of themselves as well as for understanding future mathematics.

Chapter 2

Connections: Looking Back and Ahead in Learning

This chapter focuses on the connections between foundational ideas of addition and subtraction and more basic ideas about number and numeration, as well as more advanced or complex topics, such as rational number concepts and procedures, algebraic thinking, measurement, and data representations. The big ideas discussed in chapter 1 encompass and extend earlier, simpler concepts while laying a reliable foundation for later, more sophisticated ideas. Big Idea 1 is important for work not only with whole numbers but also with other number forms, such as fractions and integers. The kinds of problems discussed in chapter 1 apply to these other numbers, supporting students' growing understanding of operations on them as well as underlying concepts of algebra (see Barnett-Clarke et al. [2010]). Big Idea 2 makes a direct connection to earlier ideas, since students put place-value concepts and the properties of addition to use in both student-generated and standard algorithms for addition and subtraction.

Big Idea 2 also connects to students' later work in mathematics. Place value and the properties of addition—first applied in whole number addition and subtraction—connect to a myriad of other mathematics concepts across the five content strands—number and operations, algebra, geometry, measurement, and data analysis and probability (NCTM 2000). The properties of addition, for example, are connected to multiplication and division and the properties of those operations. Algebraic thinking has a strong connection to Big Idea 2 because the properties of addition are generalized rules, and when students are analyzing general properties that work in addition, they are using algebraic thinking. Measurement and data representations make use of place value and properties of addition.

The discussion that follows first considers connections related to Big Idea 1—specifically, ideas associated with part-part-whole and part-whole relationships, continuous and discrete models, and measurement. The discussion then turns to connections to Big Idea 2,

 Big Idea 1

Addition and subtraction are used to represent and solve many different kinds of problems.

 Big Idea 2

The mathematical foundations for understanding computational procedures for addition and subtraction of whole numbers are the properties of addition and place value.

49

including ideas related to place value, multiplication and division, algebra, measurement (revisited), and data representation.

Links to Big Idea 1: Extending Models beyond Whole Numbers

The extension of whole number ideas to rational numbers is detailed in *Developing Essential Understanding of Rational Numbers for Teaching Mathematics in Grades 3–5* (Barnett-Clarke et al. 2010).

When children first learn to add and subtract, they typically confine themselves to whole numbers or counting numbers. However, they gradually discover the need to extend the set of numbers that they are using to include positive fractions, and they soon discover the need to add and subtract with these numbers. Applying familiar representations of addition and subtraction on whole numbers can help them understand these operations on new numbers.

Part-part-whole and part-whole

A major concept, captured in Essential Understanding 1c, is that addition and subtraction problems can be described by using the language of part-part-whole number relations or sentences. Interpreting addition and subtraction in terms of part-part-whole relationships allows these operations to be extended to other forms of numbers, such as fractions, decimals, and percents. For example, the "action" and "no-action" types of problems discussed in chapter 1 also apply to rational numbers. The problem $7/8 - 3/4$, for example, can be thought of as the action of "taking away" $3/4$ from $7/8$, or of comparing the size (measure) of the two fractions. Giving students problems with different parts missing (the *start*, *action*, or *result*), while less common in teaching rational numbers than in teaching whole numbers, is important in developing students' fluency in computing with rational numbers. Encountering these varieties of problems with whole number addition and subtraction can support students' skill in working with fractions. Although it is generally accepted that understanding whole numbers is important to learning fractions, the ways in which students' understanding of whole numbers can support or inhibit their understanding of fractions are not well understood (Mack 1995; Smith 1995; Van de Walle, Karp, and Bay-Williams 2010). There are similarities *and* differences between whole numbers and fractions that are important to know and to share with students. One of those is the conceptual relationship between "part-part-whole," used to describe addition and subtraction, and "part-whole," used to describe ratios and fractions.

➡ Essential Understanding 1c

Many different problem situations can be represented by part-part-whole relationships and addition or subtraction.

As discussed in chapter 1, the notion of "part-part-whole" is that two parts are combined to compose a whole, meaning that the two parts add to make the whole. Thus, the part-part-whole interpretation indicates an equation. "Part-whole," by contrast, does not indicate an equation but the relationship of one part to the whole.

Yet, in both part-part-whole and part-whole representations, the part is being related to the whole. The fraction $5/12$, for example, can represent 5 inches, or part of 12 inches, a whole foot.

Figure 2.1 illustrates this use of $5/12$. Notice that the figure shows a "missing" part of the foot (7 inches), so we can relate the situation to the idea of part-part-whole if we think about the whole foot as 5 in. + □ in. = 12 in. Observe how closely the idea of part-part-whole is to that of part-whole, with an important distinction: the second "part" is not apparent in the notation $5/12$. With fractions, part-whole can represent part of a *region*, part of a *group* (like a classroom of people), or part of a *length* (like the example above). Just as addition has other representations besides part-part-whole, so fractions have other interpretations besides part-whole, but part-whole is the most common one in textbooks and in classrooms. Likewise, just as students benefit from many types of addition story problems, so they would understand fractions better if instruction placed more emphasis on other representations of them (Clarke, Roche, and Mitchell 2008; Siebert and Gaskin 2006; Lamon 1999).

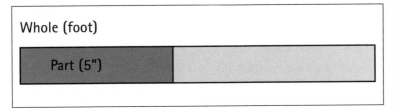

Fig. 2.1. Representing 5/12 as part of a foot

In a decimal quantity such as 0.3, the part-whole relationship is less apparent in the way in which the number is expressed because the whole is not explicit in the notation, but the expression 0.3 still means 3 parts of 10. In the case of 0.003, the meaning is 3 parts of the whole of 1,000. Similarly, 3% specifically means 3 parts out of 100, but it implies that no matter what the size of the whole, the part is just $3/100$ of that whole.

Two points are important here. First, the part-part-whole representation for addition can serve as a foundation on which to build part-whole concepts related to fractions, as long we make the distinction between the two. The example above of 5 inches can help to illustrate how to make this connection. Any time that we give students a problem with a fractional amount (for example, $1/8$ of mile walked), representing it as a part-part-whole relationship (1 + 7 = 8) and connecting it to the parts ($1/8$ part of mile walked and $7/8$ part of mile not walked), can help students to see the connection.

The second point is equally important: using a part-whole interpretation is a way for students to see the connections among fractions, decimals, and percents. Students who focus on the big

The meanings of fractions, decimals, and percents and ways to express rational numbers are elaborated in *Developing Essential Understanding of Rational Numbers for Teaching Mathematics in Grades 3–5* (Barnett-Clarke et al. 2010).

question, "What part of the whole is □?" are able to see that "4 parts out of 5" is equivalent to "8 parts out of 10" and "80 parts out of 100." They see that these quantities are proportionally the same and understand that they can write this part-whole relationship in a variety of ways, including $^4/_5$, $^8/_{10}$, 0.8, 0.80 and 80%. Understanding proportionality and multiple representations are essential to becoming mathematically proficient.

Discrete and continuous models

The terms *continuous* and *discrete* may bring back memories of high school or college algebra discussions of functions, but the discussion here focuses on models or types of manipulative or visual aids used to illustrate addition and subtraction concepts. When applied to a model or manipulative aid, *discrete* means that it cannot represent quantities between two numbers. Connecting cubes, for example, are typically used with each cube representing one whole. It is therefore difficult for students to see or represent quantities such as $12^1/_2$ or 123.6 with connecting cubes. Counters, such as those used on the ten-frames in chapter 1, are also discrete models. The hundreds chart, even though it shows numbers in a line, is another discrete model, though the hundreds chart lends itself to addition and subtraction methods that mirror the use of the number line.

Discrete models allow grouping and regrouping. These models, therefore, lend themselves to situations in which students are working with place value and are decomposing and recomposing numbers to add more efficiently. For example, in the equation 18 + 32, students may use base-ten blocks to represent each number, and then group the 8 and 2 to make another ten, arriving at the answer of 5 tens, or 50. Mentally, with a foundation in discrete models, students may decompose the problem into 18 + 30 + 2 and recompose this sum to 20 + 30 to get 50. Through the model, students may see the tens, add those, and then see the ones, add those, and then combine to reach a solution. A contrasting approach with a continuous model is discussed below.

The regrouping that discrete models allow can extend to modeling decimals and decimal operations. In a set of base-ten blocks, if the hundreds-block, or "flat" (10-by-10-by-1 piece), is designated to represent one whole, then each tens block, or "rod" (10-by-1-by-1 piece) becomes one-tenth (0.1) of the whole, and each ones-block, or "unit" cube (or "single"; 1-by-1-by-1 piece), becomes one-hundredth (0.01) of the whole. Alternatively, a thousands cube (10-by-10-by-10 piece) could also be designated as one whole, in which case each little unit cube would become one-thousandth (0.001) of the whole. Even though the unit can be assigned to represent hundredths or thousandths, there are still quantities between hundredths or thousandths that cannot be represented by the model, and hence, the model is still discrete.

Continuous models, by contrast, offer students the chance to "see" the magnitudes of numbers in a different way—along a continuum. By using a number line, students can see where 42 lies in relation to 50, or 100, or 1000, for example. Continuous models—in particular, the number line—are useful for all types of numbers. The number line is the model that is commonly used with integers, with the line extending to the left of zero as well as to the right. Students can use it with fractions and for adding fractions with like or unlike denominators (see the example in the next section, on measurement). By working with the number line, they can develop ideas about decimals much as they develop ideas about whole numbers. For example, the "jumps" on the number line in figure 2.2 show 18 + 32, but we can adapt the number line so that the tick marks represent ones and are relabeled 0 through 10, and then the jumps illustrate the problem 1.8 + 3.2.

For additional discussion of continuous and discrete quantities and models, see *Developing Essential Understanding of Number and Numeration for Teaching Mathematics in Prekindergarten–Grade 2* (Dougherty et al. 2010).

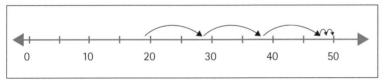

Fig. 2.2. "Jumps" marked by a student to illustrate adding 18 + 32 by first counting by 10s and then adding 2 ones

As in adding and subtracting whole numbers, the use of both discrete and continuous models in adding and subtracting fractions and decimals allows students to use different strategies (e.g., decomposing and jumping), and develop a deeper understanding of the operations.

Measurement

One of the most important advantages of the number line model is that it allows students to integrate concepts related to measurement. Measurement is closely linked to addition and subtraction, a fact that becomes very obvious when students are using number line models for learning to perform these operations. This chapter discusses measurement twice. It addresses the topic here, in relation to Big Idea 1, because measurement is often connected to comparisons, and comparison situations are one of the problem types described in connection with Essential Understanding *1c* (see example 10, for example, on p. 23). Later, the chapter revisits measurement, in relation to Big Idea 2, as one of the content connections for addition and subtraction.

A ruler with subdivisions for halves (and eventually fourths and eighths) can help young children extend their knowledge of

Essential
Understanding 1c

Many different problem situations can be represented by part-part-whole relationships and addition or subtraction.

adding and subtracting whole numbers to rational numbers. For example, if students use a ruler that has half inches marked, as shown in figure 2.3, they can use a "jump" strategy to add sums not only like 2 + 4, but also like $2^1/_2 + 1$, and so on, seeing and conceptualizing the lengths of a $^1/_2$-unit jump and a whole-unit jump.

Fig. 2.3. A ruler with halves marked can be useful for linking whole number addition and subtraction to addition and subtraction of fractions.

A foundation for future mathematics

Big Idea 1

Addition and subtraction are used to represent and solve many different kinds of problems.

In summary, Big Idea 1, which recognizes the usefulness of addition and subtraction in representing and solving many kinds of problems, provides a foundation for an enormous amount of mathematics. Students' concepts of addition and subtraction as combining or taking away ("action") or comparing ("no action") can be developed and reinforced through the use of different discrete and continuous models. Teaching whole number addition and subtraction with both kinds of models is very important, since both are used to represent a range of topics, including operations with decimals, fractions, and integers, as well as measurement concepts.

Links to Big Idea 2: Place Value and Properties across Content

Big Idea 2

The mathematical foundations for understanding computational procedures for addition and subtraction of whole numbers are the properties of addition and place value.

Big Idea 2 focuses on place value and the properties of addition as fundamental to procedures for adding and subtracting whole numbers. The connection of these computational procedures to properties and place value must be made explicit in teaching. By doing so, instruction builds connections and foundations for other mathematical ideas.

Place value

The discussion of place value that follows is brief, since the relationship between place value and addition and subtraction is an integral part of Big Idea 2 and has been discussed at length in chapter 1. However, place value itself is at the heart of many of the connections to other content areas and thus merits examination here as well.

The notion of grouping by tens is the basis of the metric system. For example, knowing that 78 is 7 tens and 8 ones helps students understand that 78 millimeters means 7 centimeters (or 70 millimeters) and 8 millimeters. It is important to distinguish thinking of 78 as 7 tens and 8 ones from writing out 78 by rote as 70 + 8. The latter may not lead to the realization that the 7 in 70 signifies 7 tens.

Multiple experiences with the "split" and "jump" strategies and discrete and continuous models reinforce the important conceptual ideas of place value. Students who have such experiences will extend their place-value concepts to include tenths, hundredths, and smaller places. As students work with decimal values, their continued use of split and jump strategies can support their understanding of decimal quantities (e.g., that 0.35 is 35 hundredths, or 3 tenths and 5 hundredths, noted symbolically as 0.3 + 0.05).

As expressed in Essential Understanding 2c, place-value concepts provide a convenient way to compose and decompose numbers to facilitate addition and subtraction computations. This is true across many curriculum topics. The notion of grouping, for example, carries over to, and connects directly with, work with measurement units, though the groupings may not be by tens. With inches and feet, for instance, the basic idea of trading is the same, but now the groups consist of units of 12 instead of 10. For example, consider the sum

3 feet 6 inches + 2 feet 10 inches.

In this case, we can add feet with feet (like tens with tens) and inches with inches (like ones with ones). When we reach 12 inches, we can make a trade (composing a unit of higher quantity) for the next higher unit—in this case, 1 foot (rather than 1 ten). The metric system, already in base 10, is very clearly linked to place value (adding like units).

This notion of adding the same-sized parts gains even more importance as the foundation for understanding the addition of fractions. Consider the fraction $1/3$. It represents a unit, three of which make a whole ($1/3 + 1/3 + 1/3$). Similarly, $1/4$ represents a different-sized unit, four of which make a whole ($1/4 + 1/4 + 1/4 + 1/4$). How can we combine $2/3$ and $1/4$? They aren't pieces or units of the same size. For us to combine them, the units (like the place values of numbers or the units in measurements) must be the same. In this case, we can use the foot as a model to help us solve the problem: $2/3$ of a foot is 8 inches, or $8/12$ of a foot, and $1/4$ of a foot is 3 inches, or $3/12$ of a foot. In our model, the sum is therefore $11/12$ of a foot; the solution to our problem is hence $11/12$. These fraction equivalences ($2/3$ and $8/12$, for example) represent *trading*—for example, we trade every one third for four twelfths—and we make these trades to obtain same-sized units for combining.

Essential Understanding 2c

Place-value concepts provide a convenient way to compose and decompose numbers to facilitate addition and subtraction computations.

When students are older, they confront mathematical situations in which they need to add like terms in expressions such as $3x + 5y + 8x + 3x$. It is their understanding of whole number addition—in particular, the importance of adding numbers of like size—tens with tens, and ones with ones—that builds the foundation for understanding that the x-terms are based on the same quantity and can be combined, but the y-terms are not necessarily based on that quantity and so cannot be combined with the x-terms. Adding functions and vectors is similarly grounded in this notion of adding like quantities.

Multiplication and division

In working with whole numbers, students encounter multiplication as repeated addition and division as repeated subtraction, so the connections of addition and subtraction with multiplication and division are embedded in these descriptions of the operations. But time and experience are necessary for those connections to become explicit to students.

Adding doubles may help students make a first connection from addition to multiplication, since it is the beginning of combining equal-sized sets. Students might explore other situations in which they are adding equal-sized sets (or groups). For example, how many pieces of gum are in 3 packs if each pack has 5 pieces of gum? Students might solve the problem by adding $5 + 5 + 5$. They might skip-count or add in their heads: 5, 10, 15. Symbolically, this process is written as 3×5, and for students to make the connection between addition and multiplication, they might attach words to this expression, such as "3 packs of 5," "3 sets of 5," or "3 groups with 5 in each group." Teachers often guide students in exploring multiplication by using an area model. In a grid that is 7 by 6, for example, a student can see 7 rows of 6 (or vice versa). The grid lets the student see the repeating rows that are all the same length, with the visual image reinforcing the connection between addition and multiplication. Understanding this connection allows students to decompose the rectangle into two smaller rectangles, one that is 7 by 5 and one that is 7 by 1, if they find it easier to calculate 7×5, and then add on 7 more.

Division is usually defined as the inverse operation of multiplication, but just as multiplication of whole numbers can be described as repeated addition, division of whole numbers can be described as repeated subtraction. Consider the following example:

> Katie was preparing treat bags for her birthday party, and she wanted to put four stickers in every bag. If Katie has 24 stickers, how many bags will she be able to fill?

Notice that this wording does not describe a sharing (which might be expressed by the question, "How can 24 stickers be equally shared among 8 bags?"). Students might solve this problem by counting down from 24 by fours (20, 16, 12, ...), recording a tally to keep track of how many bags are getting filled (equal subtraction), or they might solve it by counting up to 24 by 4s (equal addition). Story problems of this type (with the number of groups unknown) are not often used in the teaching of division, so students frequently solve such problems by using repeated subtraction, without making the connection to division.

Algorithms for multiplication and division are grounded in the properties of addition and multiplication—namely, the commutative, associative, and distributive properties. For example, by applying place-value concepts, students can think of the problem 4×18 as $4 \times (10 + 8)$. By applying the distributive property, they can see that this means that there are four 10s and four 8s. Students can compute these mentally and then add 40 and 32 to get 72. Students can also solve many division problems mentally by applying properties. For example, they might solve $132 \div 12$ by thinking about the problem in the following way: $132 \div 12 = (120 + 12) \div 12 = (10 + 1) = 11$. Or they might use a "counting up" strategy, thinking that ten 12s are 120 and one more 12 makes 132, so counting up to eleven 12s will get them to 132. As stated in Essential Understanding 2a, knowing the relationships among addition, subtraction, multiplication, and division enable students to solve problems such as this one flexibly, by using any one or more of the other operations.

Addition and subtraction are also implicit in the standard algorithm for division. Consider the example of $134 \div 8$, shown in the table in figure 2.4. Column 1 shows the steps in the standard algorithm, and column 2 shows the thought process involved at each step. Representing these steps as equations, as in column 3, illustrates the use of the relevant properties. In division, the process is carried out in partial products.

Learning basic multiplication facts is connected to addition place value and properties. Consider the basic fact 6×7. This is a challenge for many students. If students understand that 6×7 means 6 groups of 7, or $7 + 7 + 7 + 7 + 7 + 7$, they can group the sevens however they like. One common way is $5 \times 7 + 7$. But that is not the only way. If a student knows that 3×7 is 21, for example, then he can interpret the problem as $(7 + 7 + 7) + (7 + 7 + 7) = (3 \times 7) + (3 \times 7) = 21 + 21 = 42$. Moving between addition expressions and multiplication expressions helps students understand the connections between the operations and use them in solving problems, eventually memorizing the facts. It is the properties of

Essential Understanding 2a

The commutative and associative properties for addition of whole numbers allow computations to be performed flexibly.

For a discussion of the properties of division, see *Developing Essential Understanding of Multiplication and Division for Teaching Mathematics in Grades 3–5* (Otto et al. 2011).

Written calculation	Accompanying thought process	Implicit equation(s), applying properties of addition and multiplication
8)̄134	How many 8s are in 130, or how many 8s are in 13?	$134 \div 8 = ((13 \times 10) + 4) \div 8$
$\begin{array}{r} 1 \\ 8\overline{)134} \\ -80 \\ \hline 54 \end{array}$	The 1 in the quotient is in the tens place, and ten 8s are 80; subtract 80 from the dividend of 134; the difference is 54.	$\begin{aligned} 134 \div 8 &= (80 + 54) \div 8 \\ &= (80 \div 8) + (54 \div 8) \\ &= 10 + (54 \div 8) \end{aligned}$
$\begin{array}{r} 16 \\ 8\overline{)134} \\ -80 \\ \hline 54 \\ -48 \\ \hline 6 \end{array}$	The 6 in the quotient is in the ones place, and six 8s are 48; subtract 48 from the difference of 54; 6 is the remainder.	$\begin{aligned} 134 \div 8 &= 10 + (54 \div 8) \\ &= 10 + ((48 + 6) \div 8) \\ &= 10 + ((48 \div 8) + (6 \div 8)) \\ &= 10 + (6 + (6 \div 8)) \\ &= (10 + 6) + (6 \div 8) \end{aligned}$

Answer: 16 R 6, 16 $3/4$, or 16.75

Fig. 2.4. Reasoning and equations representing properties of addition and division in the standard algorithm for division in the case of $134 \div 8$

➡ Essential
Understanding 2d

Properties of addition are central in justifying the correctness of computational algorithms.

➡ Essential
Understanding 1b

Subtraction has an inverse relationship with addition.

addition that justify the correctness of invented and standard algorithms (Essential Understanding 2d).

Algebra

The role of algebra in the primary grades should go far beyond occasional appearances in repeating and growing patterns. In fact, to switch metaphors, we might describe algebra as a lens for magnifying relationships and properties of number and operations. That is certainly true in the case of addition and subtraction. Recall that Essential Understanding 1b expresses the idea that subtraction is the inverse of addition, an idea introduced in chapter 1 by an example, part of which is shown in figure 2.5.

The table and the function machine illustrate not only the reversibility of addition and subtraction (Essential Understanding 1b), but also the *generalizability* of number. In the table, students can see that the relationship from the input to the output is "plus 2," and that the relationship from the output to the input is "minus 2." Once students have recognized the rule, they can use examples of larger numbers—for example, 100 or 500. As students see that the rule always applies, they recognize that it applies *in general*, not just in certain cases. In other words, they realize that the rule

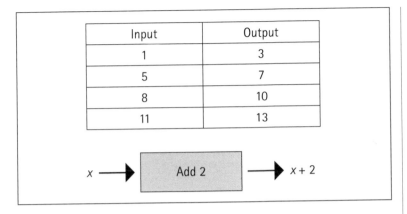

Input	Output
1	3
5	7
8	10
11	13

Fig. 2.5. Add 2 as an addition operation

is a *generalization*. When students generalize a series of numerical examples, they are doing algebra. Suppose that students are working with the following problem set:

$$14 - 9 = \quad 12 - 9 = \quad 15 - 9 = \quad 18 - 9 =$$

Consider how using an algebraic lens on their thinking can support students' strategies for finding the differences and later memorizing these facts.

Students may use a range of tools and strategies to solve these problems. A hundreds chart is a nice tool for visualizing how the initial value and the result relate to each other. Instead of stopping after students obtain correct answers, the conversation needs to continue to include patterns that students notice, testing those ideas. These generalizations are algebraic thinking, and such reasoning is central to learning addition and subtraction well. Students might notice that in all cases they could have used $-10 + 1$ instead of -9. The discussion of this set of problems should include (at least) two important questions:

1. Does this $-10 + 1$ "rule" apply to all of our examples?

2. Will this $-10 + 1$ "rule" apply to any "minus 9" problem, or when will this rule work?

In this case, this rule always works, and symbolically it is written as $n - 9 = n - 10 + 1$.

The notion of decomposing and composing numbers runs throughout this book, and it has its roots in algebraic thinking. If 7 is the whole, for example, in a part-part-whole situation, what could the parts be? They could be 3 and 4, 1 and 6, and so on. Students can be asked to find all the (whole number) ways of decomposing 7. A popular context (Yackel 1997) involves monkeys in trees: How can students distribute 7 monkeys in two trees. The more challenging question that students must ultimately answer is, "How do we know that we have found all of the solutions?" Students may make a list or table like that in figure 2.6 to find all the ways.

Big tree	Little tree
5	2
6	1
4	3

Fig. 2.6. Table representing ways to distribute 7 monkeys in 2 trees

As students are looking for patterns in the table and trying to notice a rule for knowing when they have found all the ways, they are generalizing, and therefore they are using algebraic thinking. Students might notice, for example, that for every digit from 0 to 7 there is one way (0 + 7, 1 + 6, ..., 7 + 0). Carpenter, Franke, and Levi (2003) report that second graders noticed that there is one more possible way than the size of the number—in this case, 8 ways.

Finally, as students explore addition and subtraction story problems, using an algebraic lens can expand the way in which they write equations. As suggested throughout chapter 1, numerical relationships in story problems can be represented in several ways— for example, as $a + \square = b$ or $b - a = \square$. The first equation shows a missing addend, an unknown quantity that students will later represent with a variable. Asking students to solve equations in algebraic form, like $5 + \square = 4 + 8$, can support their understanding of the equals sign, give them experience with sums, and help them understand number relationships. In the case of $5 + \square = 4 + 8$, they can apply the idea that if the first addend on the left (5) is one less than the first addend on the right (4), then the second addend on the left must be one more than the second addend on the right (8). As the discussion in this section has illustrated, addition and subtraction do not simply provide foundations for algebra (or vice versa); these arithmetic operations and algebra are intimately interrelated and mutually support learning in both domains.

Chapter 3 offers a full discussion of the important issue of developing understanding of the equals sign by working with, and developing concepts related to, addition and subtraction.

Measurement concepts

Measurement, like algebra, is very closely linked to concepts of addition and subtraction. The related concepts that students explore in the measurement strand include measuring time, weight, and length; comparing the lengths of two objects (subtraction); combining lengths in various contexts, including perimeter problems; and combining areas of shapes (or angle measures). The following examples show how close the connection is between measurement and addition and subtraction:

Example 1: Your brother says that you can play the video game in 45 minutes. You look at the clock and see that it is 5:07. What time should you come back for your turn to play?

Example 2: Suppose that a snake grows from 5 inches at birth to 17 inches as an adult. How many inches does the snake grow from birth to adulthood?

Example 3: Suppose that an aerial view of one section of fence is shown on the left below and one patio tile is shown on the right:

How many fence sections do you need to enclose the patio shown below? If the whole patio is also going to be covered in tiles, how many will fit in the space?

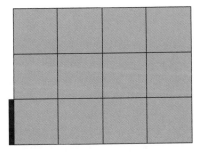

Students' probable approaches to these problems are discussed below.

Students might solve the problem in example 1 in a variety of ways. One would be to recognize that they are adding 45 and 7 and do so by first splitting 7 and then adding 45 + 5 + 2 = 52 minutes, to get 5:52 as the time to come back for the video game. Or students might solve this problem by counting on, or jumping, four times by 10 and then one time by 5: 5:17, 5:27, 5:37, 5:47, 5:52. The first strategy shows that the students recognize that they are adding minutes and are thinking about minutes as counting objects. The second strategy is aligned conceptually with the disk of an analog clock, which is really a segment (from 0 to 12) of the number line curved into a circle.

Example 2 presents a comparison situation. Students are comparing the snake's length as an adult to its length at birth. They could make physical comparisons with drawings or with a manipulative like connecting cubes or Cuisenaire rods (in fig. 2.7, the variable *b* represents birth length, and the variable *a* represents adult length). Students are likely to use a "counting up" strategy, because of the context and the way in which the story problem is written. However, they may also count down from 17. Encouraging multiple ways to think about the problem is important in deepening student understanding of addition and subtraction.

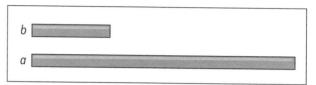

Fig. 2.7. Comparing measurements as a context for subtraction

Although area and perimeter may be content goals for later grades, example 3 cleverly builds ideas that will support later learning of these concepts while simply providing a context for adding. Students can find the perimeter by counting fence sections one by one (possibly acting out the process with materials), or they can find how many fence lengths are needed for each side and then add those four quantities together. They can also determine how many tiles will fit in the space by counting individual tiles (possibly using color tiles to make a model of the space) or by seeing that there are three rows of 4 and using repeated addition. Young children can solve more challenging problems of the same type. For example, students can cut out nets for open boxes and apply their addition skills to figure out how many squares are on their surface and how many cubes might fill them. For young students, these measurement problems provide an interesting context for measuring and adding— not for finding surface area and volume formally. Students can employ a hundreds chart, a number line, or base-ten blocks in figuring out the total quantity as the numbers get larger.

Measurement and addition and subtraction are synergistic. Measurement provides an interesting context for adding and subtracting, and using such contexts supports students' understanding of both measurement and the operations.

Data representation

Strategies for analyzing and displaying data are grounded in concepts of addition and subtraction. When young children create bar graphs, they should analyze what they see. They can compare data (by considering such questions as, "How many more people liked horses than liked sheep?") or combine data (by considering questions like, "How many people picked a farm animal?").

For additional discussion of quantity and measure, see *Developing Essential Understanding of Number and Numeration for Teaching Mathematics in Prekindergarten–Grade 2* (Dougherty et al. 2010).

Venn diagrams give young children a different way to look at and compare data. Consider figure 2.8, which shows data gathered by a group of 18 first graders about themselves. Initially, the teacher might pose counting questions for the students to answer by examining the data, such as the following:

- "How many of you have both a sister and a brother?" (3)
- "How many students have sisters?" (7; requires combining the number of tokens in the overlap of the "sister" and "brother" circles with the number of tokens in the "sister only" section of the "sister" circle)
- "How many have brothers?" (10)
- "How many of you have a brother or a sister?" (14)

The "or" question is the most difficult and relates to an important property for addition—that quantities to be added cannot overlap (see example 13 in chapter 1; p. 25). Students will debate whether the answer is 17, obtained by adding 7 + 10, or 14, obtained by counting the total inside the two circles of the Venn diagram: 7 + 3 + 4. Having students stand up if they have a brother or sister (or both) will illustrate that the correct response is 14. It is important for teachers to show that the other answer counts 3 students (those in the overlap) twice. This notion of distinct (or discrete) groups becomes more complicated in middle and high school, when students use formulas for counting. Early experiences such as this one can provide a conceptual foundation.

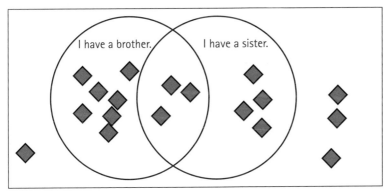

Fig. 2.8. Data gathered from a group of grade 1 students

A link to later topics

In summary, Big Idea 2, which recognizes place value and the addition properties as fundamental to computational procedures for addition and subtraction, relates to many critical topics in the mathematics curriculum. By working with contexts in algebra, measurement, and data, and by making explicit connections to place

Big Idea 2

The mathematical foundations for understanding computational procedures for addition and subtraction of whole numbers are the properties of addition and place value.

value, students gain a deep understanding of addition and subtraction and are ready to explore more advanced topics that depend on this understanding.

Conclusion

A quick glance back at the headings in this chapter gives an indication of the vast importance of addition and subtraction to other mathematics. The connections described above collectively raise two important points.

First, having a strong and deep conceptual understanding is what allows addition and subtraction to serve as foundational topics. For example, if students do not understand the part-part-whole nature of addition and subtraction, they cannot use it as a foundation to constructing the notion of part-whole. Similarly, students who frequently use discrete and continuous models in addition and subtraction can apply them to multiplication and division and to operations with fractions and decimals. As described above, students can apply their whole number knowledge of jumps to solving decimal operations, but only if they have had such experiences with whole number addition and subtraction.

Second, conceptual understanding and procedural proficiency are intertwined, as described in *Adding It Up* (Kilpatrick, Swafford, and Findell 2001). Connecting place value to addition and subtraction, connecting addition to multiplication (and division), and seeing the generalizability of procedures (algebra)—to name just a few of the connections described above—require both conceptual understanding of and procedural proficiency with addition and subtraction. In other words, developing a sound understanding of addition and subtraction supports *and* is supported by place-value concepts, algebraic thinking, and ideas from other topics. When it comes to computation, proficiency in addition and subtraction of whole numbers is essential to developing mathematical proficiency with other number forms, such as fractions, decimals, and integers.

Chapter 3

Challenges: Learning, Teaching, and Assessing

Consider the following dialogue between two collaborating teachers as they focus intently on which problem to use to begin their students' inquiry into subtracting with numbers in the teens. While these teachers debate the merits of each combination, they are integrating their knowledge of content with their pedagogical knowledge—their understanding and experience of how students learn—to make the best professional decisions to enhance their students' learning of subtraction.

Let's listen in as these two teachers plan. Ms. Wilson and Mr. Lee are negotiating about the best way to get students to think about the inverse relationship of subtraction and addition. They have made two decisions: (1) they want to use a context involving a girl who has seven trading cards and needs to have a larger quantity to complete the full set, and (2) they intend to ask, "How many more does she need to collect?" They are debating whether to use $13 - 7 = \square$ or $12 - 7 = \square$ as the basis for the lesson. They want to create opportunities for students to build on their prior knowledge of place value and possibly identify and explore the use of 10 as a benchmark in finding differences. What Ms. Wilson and Mr. Lee have to say may surprise you:

Ms. Wilson: I think we need to set up the problem so that students will investigate the idea that the difference between the two numbers can be found by adding up in jumps and using 10 as a landmark. This is the same reasoning that will help them when working with larger numbers.

Mr. Lee: I agree, but I'm thinking that 12 might be a better number to use, since they're more likely to know the combination of 3 plus 2 than 3 plus 3. At this point, I want to keep the combinations as easy as possible, so the students can focus on understanding the process of subtraction rather than try to figure out more challenging combinations. We can add the more complex numbers later when the subtraction process seems to be firmly in place. I also think it makes sense contextually that a full collection of trading cards would be a dozen. That might be easier to relate to.

Ms. Wilson: I see your point, but I actually think that the students will be more familiar with the "doubles" combination of 3 plus 3, and that the symmetry, or evenness, of the difference in relation to 10 will help them identify this way of thinking about the problem. I can see them grabbing their ten-frames here to visualize the comparison between 13 and 7. I think using 13 in the problem is more likely to be a good starting point for their investigation. We will also need to connect this to the corresponding missing addend problem of 7 + □ = 13.

Setting the Stage for Addition and Subtraction

Like Ms. Wilson and Mr. Lee, as you set the stage for your students' learning of addition and subtraction, you must consider their prior knowledge carefully. You must work purposefully and strategically to bridge the gap between the mathematics experiences that students bring from home and the approaches to learning mathematics that you present in school. The foundation that you establish for additive thinking will influence your students' reasoning about operations and their formation of the underlying mathematics concepts that will develop into multiplicative thinking and provide the scaffolding for more advanced learning.

When you are considering instructional tasks related to addition and subtraction, your students' understanding should be a key focus. Specifically, it is valuable to consider students' mastery of these topics in light of the five strands of mathematical proficiency identified by Kilpatrick, Swafford, and Findell (2001): conceptual *understanding*, *computing* (procedural fluency), *applying* (strategic competency), adaptive *reasoning*, and *engaging* (productive disposition). The acronym formed by this list of the elements of proficiency—UCARE—is an easy way to remember these components (National Research Council 2002). These five elements of proficiency are interwoven through the exploration that follows of the pedagogical considerations that are related to addition and subtraction in the primary grades.

Representations and Models

The concrete–semi-concrete–abstract (CSA) teaching sequence (also known as the concrete-representational-abstract [CRA] instructional approach) has been used in mathematics education in a variety of forms for years (Heddens 1964; Witzel, Mercer, and Miller 2003). This model reflects a continuum from concrete representations and models to semi-concrete representations and images to symbols and abstraction. The thinking that students develop while exploring concrete representations often allows them to connect their ideas to abstract representations. Built into this approach is a cyclic return through the components of the model as students learn new concepts in ways that link to previous ideas. The CSA model emphasizes conceptual understanding as a vehicle for developing mathematical proficiency. As students' thinking is developing through each representation, this process supports connections and moves students to full understanding and mastery of addition and subtraction.

It is important to recognize that the manipulative materials that you choose to support your learning goals for teaching addition and subtraction will influence your students' thinking and shape the conceptual models that the students develop. Whether you select one-inch cubes, plastic counters, bundles of sticks, base-ten blocks, hundreds charts, or number lines, these materials will provide your students with rich visual representations for concepts and lay a foundation for more abstract thinking. Although manipulative materials can be powerful tools to support students' thinking, using them in effective ways is a complex task (Leinhardt et al. 1991). As Ma (1999) states, "A good vehicle, however, does not guarantee the right destination. The direction that students go with manipulative materials depends largely on the steering of their teacher" (p. 5). For example, a teacher may use a large cube from a set of base-ten materials to represent 1000 in an addition problem, but some young students will see the squares etched on the six faces of the cube and incorrectly assume the cube represents 600. The materials are often more abstract than we think; as adults, we have made the conceptual leap.

As an example of the influence of manipulative materials, consider what frequently happens when we wish to move students beyond the use of a counting strategy to prepare them to deal effectively with larger numbers and we make the strategic choice of a number line or empty number line as a model. Faced with this particular representation, students are likely to use a "jump" strategy, discussed in chapter 2 and again below, as they "chunk" numbers and jump to positions on the line.

→ Essential
Understanding 1c

*Many different
problem situations
can be represented
by part-part-whole
relationships and
addition or
subtraction.*

The Structure of Word Problems

As described in chapter 1 in relation to Essential Understanding 1c, the structures of word problems in addition and subtraction have been analyzed and frequently categorized into two major groups: (1) part-part-whole, representing "join," "combine," "separate," or "take away" situations, as well as some situations with no action; and (2) comparison, representing situations focusing on the difference of two sets, finding the small set, or finding the large set. What does this mean for teaching and assessment? The story problems that you select should cover the spectrum of these situations; if they do not, your students will struggle to find ways to solve the types of problems that you omit. For example, if students encounter only "take away" stories—the type most commonly presented by teachers for subtraction—they won't recognize that a problem comparing two students' heights presents a subtraction situation. In fact, comparison problems are less common in textbooks and in classrooms and therefore are more likely to be missed by students on assessments. In particular, comparison problems with the reference amount unknown are the most difficult problems for students to decipher and the least often highlighted in practice. Two examples of these problems follow:

1. Courtenay has 10 music CDs. All of her CDs are either country music or hip-hop. She has 4 more country CDs than hip-hop CDs. How many country CDs and how many hip-hop CDs does Courtenay have?

2. Morgan has 7 apples and oranges in the refrigerator for healthy snacks. He has 3 more apples than oranges. How many apples and how many oranges does Morgan have?

It is important to use a variety of problem structures so that students have opportunities to engage in discussions of models and situations that arise in real-world settings.

Special education researchers suggest that students with disabilities be explicitly taught these underlying structures so that they can identify the features of the situations and judge when to use addition or subtraction (Fuchs et al. 2008; Xin, Jitendra, and Deatline-Buchman 2005). Fuchs and colleagues (2008) recommend that teachers help students understand how to separate important information in the problem from superficial details as well as how to apply these structures to problems presented in tables or charts. When these students are then exposed to novel problems, this emphasis on problem structure and relevant features will assist them in generalizing from similar problems on which they have practiced (Fuchs et al. 2004).

Furthermore, students encounter challenges as they try to translate word problems into representations or number sentences. For example, consider the following problem:

> Liza had 8 apples. Then she picked 5 more apples. How many apples did she have then? Can you show how you thought about this problem?

This problem can be translated directly from the context as $8 + 5 = \square$. Contrast that to the situation in the following story problem:

> Ermando has 12 cents. He needs 25 cents to buy the trading card that he wants. How much more money does he need?

This problem may be translated as $12 + \square = 25$, aligned to the meaning of the story, or it may be translated as $25 - 12 = \square$. It could be translated as $25 - \square = 12$. As Essential Understanding 1d suggests, a story problem structured in this way is likely to lead to a range of equations, an outcome that provides an excellent opportunity for deepening students' understanding of addition and subtraction. Encouraging these different ways to write an equation, discussing their equivalence, and justifying how each equation correctly represents the situation help to build the robust understanding captured in the UCARE components of proficiency.

Students can navigate more capably through the task of translating from language statements to symbolic representations in number sentences if they work explicitly and strategically (Lochhead and Mestre 1988). Using information expressed in word problems, students often mistakenly move from left to right, word by word, as they try to craft an equation. This decoding of language and conversion of it to numerical symbols are particularly challenging for English language learners (ELLs). Again, the underpinnings of algebraic thinking come into play. Lochhead and Mestre (p. 133) suggest a three-step process, involving students' *qualitative*, *quantitative*, and *conceptual* understanding, from which we can construct the following model:

1. **Explore students' qualitative understanding.** As you probe your students' grasp of the problem, be on the lookout for foundations of understanding on which you can build by asking students how well they comprehend the problem (you can ask older students to restate it in their own words). Then try to separate the students' difficulties in reading or interpreting (especially in the case of ELLs) from any possible mathematical confusion that they are experiencing, remembering that sometimes both will come into play. A question that you might ask is, "Are there more _____ or

Essential Understanding 1d

Part-part-whole relationships can be expressed using number sentences like $a + b = c$ or $c - b = a$, where a and b are the parts and c is the whole.

_____?" Or "Are the 8 apples *part* of the whole amount of apples or *all* of them?"

2. **Explore students' quantitative understanding.** If students struggle in expressing a basic understanding of the problem situation, then you might move to a simpler case of the same problem to see how well they understand the numerical relationships. A question that you might ask is, "Suppose there are 10 _____; how many more _____ would you need?"

3. **Explore students' conceptual understanding.** Finally, as students write their equations, pose questions to probe their understanding. A question that you might ask is, "What would happen if we changed the number from _____ to _____?" Or you might ask, "What would happen if they were getting more _____ instead of losing _____?" Another possible question might be, "If you know you need _____, and you have _____, how can you find the other part?" Or, to spark student-to-student discussion, you might direct attention to a particular student's work and say, "Look, _____ wrote an equation that is different from yours; can both equations be right?"

Note that in all three steps, you are not giving the "right" answer but are helping your students move to a position where they are confident of their approach. This shift refocuses the conversation on the relationships among the numbers and keeps the students' thinking from stopping once class members share the right answer.

Key words: A failed shortcut

Research indicates that it is wise to avoid teaching students to use a key-word strategy in solving word problems. When young students are taught to use such a strategy, they often misinterpret words and shift from a focus on analyzing the problem and sense making to a mechanical and potentially inappropriate reliance on key words (Kenney et al. 2005). If students are taught to search for key words to unravel word problems, they are likely to try to decipher all such problems in that fashion, without attempting to interpret them meaningfully in context. Not surprisingly, this can lead to misinterpretation and confusion. For example, consider the word *more* in the following two situations:

1. Jessica had 9 pennies. She spent 5 pennies, and then she lost 1 more. Now how many does she have?

2. MacKenna has 4 cartons of yogurt. How many more does she need so that she can have 1 carton of yogurt for breakfast for 7 days?

Students who think that *more* is the key word and means *add* will mistakenly do just that. Other words and phrases, such as *altogether* and *in all*, may also be misleading if students believe that they always suggest a single path to addition. The same can be said for *left* and *fewer* if students believe that they invariably suggest subtraction. More important, many problems do not have key words, leaving students who have learned only a key-word approach without a way of determining a solution strategy. Two-step problems are almost impossible to solve through a key-word approach. Unfortunately, this is an approach that teachers often use with students who are struggling, and such students are least likely to be able to sort out the nuanced meanings of these words in word problems.

Reasoning about actions and relationships

Clearly, key-word shortcuts create misunderstandings for students that impede their progress toward mathematical proficiency, as reflected in the UCARE components. Understanding word problems is much more complex than understanding the words and their arrangement. Instead, it involves understanding the quantities and their explicit and implicit relationships, framed within the context of the story. Removing the story and the nuance from a word problem (to make the problem more accessible to students) actually removes the student from the problem.

What is best to emphasize in solving word problems is investigating the actions and incorporating reasoning rather than pulling out specific words. The very same words may be used in different types of problems (for example, *in all* may appear in addition, subtraction, multiplication, or division problems). Instead, we want to give students experiences with many different contexts and types of word problems, using different ways of indicating arithmetical actions. One of the best approaches is to have young students dictate or write their own word problems. You can even indicate what operation you would like for them to highlight. As students begin to craft the problems, they see the components and language required to develop addition or subtraction situations, and you will be able to assess their understanding.

The Structure of Equations

Research shows that the equals sign—in particular, *where it occurs* in an equation—creates consistent problems for students (RAND Mathematics Study Panel 2003). Even some older students—and adults—harbor confusion about the meaning of the equals sign and how number sentences can be written. Therefore, a diagnostic interview might be an appropriate assessment tool. A diagnostic interview is a formative assessment (usually one-on-one) that can provide in-depth information about a student's knowledge and conceptual strategies. Consider the following interchange between a teacher and a young student during a diagnostic interview regarding a series of addition equations:

Ms. Klumb: [*Pointing to the problem shown in fig. 3.1, 8 + 4 =* ☐ *+ 5*] Look at the first problem. What number do you think belongs in the box?

Aniessa: Twelve [*writing 12, as shown in fig. 3.1*].

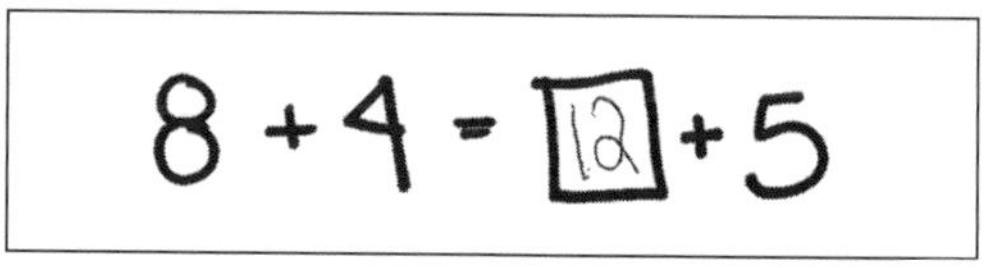

Fig. 3.1. A student's confusion, as exhibited in a missing addend problem

Ms. Klumb: How did you get that answer?

Aniessa: I started with the 8; then I counted 4 more to get 12. It's easier to start with the bigger number.

Ms. Klumb: [Pointing to the right side of the equals sign] Why does the equation show to add 5 more?

Aniessa: I need to add 5 more?

Ms. Klumb: What does the equals sign stand for?

Aniessa: It means, "What is the answer?" So, when you see an equals sign, you put the answer. That's why I wrote 12.

Ms. Klumb: Let's try the next problem [*shows the problem in fig. 3.2,* ☐ *+ 5 = 5 + 8*]. What number belongs in the box?

Aniessa: [*Writing 0 as shown in fig. 3.2*] This is easy, 0 plus 5 equals 5.

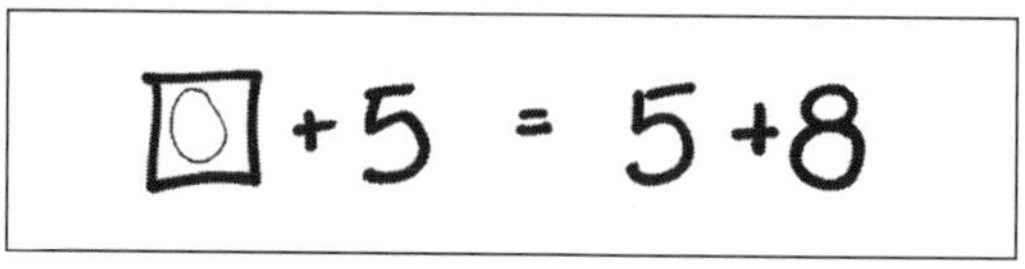

Fig. 3.2. A student's confusion about the equals sign

Ms. Klumb: [*Pointing to the right side of the equals sign*] Why does it show + 8?

Aniessa: When the equals sign goes behind the last number, then you write the answer after it.

As Aniessa's responses suggest, if you ask a student what the equals symbol means, you are likely to get an answer that it means something like "makes." In her conversation with Ms. Klumb, Aniessa said that the equals sign means "What is the answer?" Aniessa's confusion was compounded by the difference in the format of the equations from the classic form of $a + b = \square$. When older students explore higher-level algebraic thinking, they learn that the equals sign means *equivalence* (and has other meanings, as well [Verschaffel, Greer, and De Corte 2007]) in the sense that both sides of the equation are "balanced." But to set the stage carefully for that knowledge, you should emphasize from the start that the equals sign is a *relational* symbol—not an *operational* symbol (like the operators $+$, $-$, $\times$, $\div$). What appears to the left of the equals symbol should be *the same as*, or *equivalent to*, what is on the right of the symbol. By using a balance as a concrete representation (either an actual balance or a virtual manipulative), students can place values on either side and explore equations. Dougherty and colleagues (2010) address the importance of the equals sign and its use in composing and decomposing whole numbers.

Developing Essential Understanding of Number and Numeration for Teaching Mathematics in Prekindergarten– Grade 2 (Dougherty et al. 2010) discusses the importance of the equals sign.

In the same way that students should recognize multiple contexts as indicating addition, they should recognize multiple ways of writing addition and subtraction equations. The notation that most students see or use initially for addition (and subtraction) is horizontal—for example, $8 + 3 = \square$. Young students rarely see equations written in a vertical manner until they reach larger (multi-digit) addition and subtraction problems, when the novelty of the vertical arrangement may add to their difficulties in performing these computations.

Even within the framework of horizontal equations, students need to see multiple ways to record situations. Seeing equations written only in the manner of $8 + 3 = \square$ (part + part = whole) or $8 - 3 = \square$ (whole − part = part) contributes to the kind of misunderstanding of the equals sign illustrated above (McNeil and Alibali 2005; Seo and Ginsburg 2003). To avoid the pitfalls of a single approach, you can use a variety of equations with the unknown quantity in multiple positions, including equations that closely match a story situation. For example, consider the following story:

> Mark wants to know the total number of coins in his pocket.
> He takes them all out and counts 5 nickels and 7 dimes and
> puts them all back. How many coins does Mark have in his
> pocket?

This situation could be recorded mathematically as $\square = 5 + 7$, to align with the situation exactly as stated in the problem. This mathematical statement is referred to as the *semantic equation* of the problems, in that it precisely matches the order of the quantities as presented in the problem. When a situation is translated into an equation that has the unknown amount and no other quantities on the right of the equals sign, the equation is known as the *computational form* (which is also the form in which it would be entered into a calculator).

Equations can be written in several ways, and students should also be familiar with equations that include an addition statement on both sides of the equals sign. For example, students should record relational statements, such as 3 + 4 = 4 + 3 or 5 = 5. Students should also justify that a statement like the following is true: 4 + 3 = 5 + 2. These number sentences indicate the sameness in quantity of the two collections on either side of the equals sign, and working with such statements will help students deepen their understanding of the equals sign, as well as of properties such as the commutative property.

Using Appropriate Language

As students in prekindergarten to grade 2 are in the process of developing their reading literacy, they are making parallel advances in their mathematical literacy. Both reading and mathematics have fundamental rules and conventions, depend on comprehension (understanding a situation in a story or a word problem), and have significant practical uses outside of the classroom. In each case, use of appropriate terminology established in the primary years sets the foundation for learning in the future.

Terminology should always support students' understanding. Vocabulary such as *borrowing* and *carrying* focuses on procedures rather than understanding. These terms have the potential to create problems for students, including the following:

1. Students think that they can arbitrarily change a number in a subtraction problem if it is too small by "borrowing" from another number.

2. Students focus on digits rather than number values.

3. Students are not reinforcing their understanding of the base-ten number system.

Language that focuses on the concept of equivalent exchanges has much more power to strengthen and support students' developing understanding. Ma (1999) suggests that the use of language related to decomposing (or composing) a unit of higher value capitalizes on the exchange of tens as a generalizable process that can be applied to a variety of problem situations. By unbundling and bundling materials, students can experience the structure of the base-ten system. For example, when considering a problem such as $42 - 27 = \square$, students should begin to think, "Can I subtract a number in the 20s from a number in the 40s," instead of, "Can I subtract 7 from 2?" Borrowing and carrying are strictly procedural steps in which a number (e.g., 1) is moved either to the left or right.

Therefore, this commonly used, well-practiced, but ultimately ineffective language, however ingrained in us, should be replaced with terminology that honors the concepts that relate to the operations.

Part of teaching addition and subtraction involves preparing students to use these operations not only on whole numbers in a part-part-whole structure but also on other numbers that they will encounter later, such as rational numbers and integers, and expressions in algebraic systems. Language such as "addition makes bigger," and "subtraction makes smaller" (considering the first number as the point of departure) gives students rules that they discover no longer hold when they encounter integers or consider equations with zero. Many students who hear the incorrect but all-too-common assertion that a calculation such as 5 – 7 "cannot be done" eventually experience cognitive dissonance when they learn about negative numbers. In the system of integers, addition can indeed "make smaller," and subtraction can indeed "make bigger." This seeming conflict can be anticipated, and language can be crafted so that students' previous knowledge can serve as a meaningful foundation for later mathematics concepts. Then the extension of addition and subtraction to other number systems can fit in the logical progression to more sophisticated understandings. This adjustment in language allows new knowledge to be assimilated rather than accommodated, to use Piaget's (1928) descriptions of learning processes. The challenge then is to teach elementary ideas about addition and subtraction without setting up conceptual conflicts with these more advanced concepts.

Student-Invented Strategies

Just as it is important to value different ways to record equations that represent story problems, so it is important to value student-invented strategies for adding and subtracting. Invented strategies are grounded in students' understanding of number (as implied in Essential Understanding 1*a*), and they support and reflect the students' mathematical proficiency (as elaborated in the five UCARE components—understanding, computing, applying, reasoning, and engaging). Introducing traditional algorithms too early may actually cause harm by initiating a misplaced focus on memorization and procedural skill at the expense of conceptual understanding (Kamii and Dominick 1998). It is best to delay the introduction of traditional algorithms in the primary years, maintaining a number orientation instead of a digit orientation. In other words, keeping students focused on the number value is much more productive in the long run than encouraging them to take one digit at a time. Research suggests that students who use invented strategies prior to any standard algorithms have a better grasp of place value and are

Essential
Understanding 1*a*

Addition and subtraction of whole numbers are based on sequential counting of whole numbers.

more flexible in using prior knowledge to extend it to novel situations and problems (Verschaffel, Greer, and De Corte 2007).

Students need to formulate their own mental or invented strategies with one-digit numbers to help them understand two-digit (or higher) addition and subtraction. Just as many adults in real-world situations use operation strategies that they were never formally taught in school, students can begin to recognize when a process or strategy is generalizable.

Three strategic approaches are likely to emerge as students begin to invent strategies for addition and subtraction. These capitalize on the students' number sense and their resulting ability to compose and decompose numbers. We can group these strategies into the categories discussed by Verschaffel, Greer, and De Corte (2007) as "split," "jump," and "compensate."

The "split" strategy emphasizes the decomposition and recomposition of numbers largely by place value. Using this approach, students might think of 54 – 32 = □ as 50 – 30 = 20 and 4 – 2 = 2, for an answer of 22. This strategy aligns well with early use of concrete base-ten materials.

The "jump" strategy is illustrated for 54 – 32 = □ in figure 3.3:

$$54 - 32 = 54 - 10 - 10 - 10 - 1 - 1 = 22$$

Like a splitting approach, jumping also allows students to decompose numbers through chunking in a "jump-up" (addition) or "jump-down" (subtraction) approach.

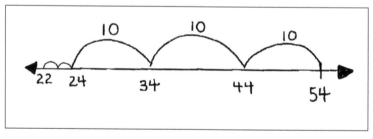

Fig. 3.3. Using the jump strategy for 54 – 32 = □

The third approach is the "compensation" strategy, which allows students to capitalize on the derived strategies for basic facts discussed in chapter 1 by creating other, more "friendly" numbers to think about and work with in solving problems. For example, to solve the addition problem 148 + 33 = □, a student might add 150 + 33 = 183 and then subtract 183 – 2 = 181.

Eventually, students will be ready to develop the standard paper-and-pencil algorithms, linked to conceptual understandings of how they work, as one way to perform addition and subtraction problems. As discussed by Verschaffel, Greer, and De Corte (2007) "standard algorithms tend to mask the underlying principles that make them work" (p. 590). Students who memorize a standard algorithm, but do not understand it, are more likely to make errors and forget how to use it (Thompson 1999). This is particularly evident when students focus on digits rather than the number (Fuson 1992). A focus on number rather than digits also honors the intersecting relationship of place value and multi-digit addition and subtraction that is emphasized in Essential Understanding 2c. For example, when thinking about a problem like $60 - 58 = \Box$, a student who is using a digits approach will carry out the algorithm by focusing first on the ones digits and then on the tens, while a student who is using a number approach will think about the numbers and mentally "see" that the difference between 58 and 60 is 2.

Essential ← Understanding 2c

Place-value concepts provide a convenient way to compose and decompose numbers to facilitate addition and subtraction computations.

There are cultural differences in the use of algorithms, and students might bring to your classroom new algorithmic approaches that reflect their prior knowledge or their families' experiences. These algorithms may also inform their current learning. For example, consider the Austrian method, discussed in chapter 1 (pp. 45–46). A popular subtraction algorithm used in many other countries and frequently known as "equal addition," or "add tens to both," this method is based on the fact that adding the same amount to the minuend and the subtrahend will not change the difference between the two. So, in finding the difference indicated by $43 - 28 = \Box$, a student using "equal addition" would mentally add 10 to the 3 in the ones column of the 43, thereby obtaining 13 in the ones column. To counteract the effect of the addition of that 10, the student would mentally add 10 to the 28 (subtrahend) and then subtract 38.

Because the "equal addition" process might not be familiar to you, you might try it to experience how it works (for an example, see fig. 1.47 and the discussion leading up to it on p. 45). You will find that this approach (among others) can serve as a wonderful context for exploring subtraction (or addition) and place value. Your heightened awareness of cultural differences in addition and subtraction algorithms can increase the comfort and sense of connectedness experienced by ELLs, who bring different knowledge to the classroom. These alternative algorithms also provide strategies that may make more sense to struggling learners who do not understand a traditional algorithm.

Assessing Evidence of Learning

Doug Clarke, a researcher in mathematics education, has teamed with colleagues to describe a powerful trajectory of growth points in the learning of addition and subtraction. This trajectory provides a framework for creating diagnostic assessment that taps students' fundamental understanding. The team's Early Numeracy Research

Project (ENRP) includes developmental assessment tasks or diagnostic interviews for each of the six "growth points" (Clarke 2001) that are summarized in the chart in figure 3.4. The growth points mark key transitions in students' growth from one type of strategy to another. These growth points are related to but not identical with the strategies named and described in this book. For example, students reach the "count-on" growth point when they move from a "counting all" strategy to a "counting on" strategy.

Growth Point	Description
Count all (two collections)	Student counts all to find the total of two collections
Count on	Student counts on from one number to find the total of two collections (eventually counting from the higher number)
Count back, count down to, count up from	Given a subtraction situation, student chooses appropriately from strategies including counting back, counting down to, and counting up from
Basic strategies (doubles, commutativity, add 10, tens facts, other known facts)	Given an addition or subtraction problem, student uses strategies such as doubles, commutativity, adding 10, tens facts, or other known facts
Derived strategies (near doubles, add 9, build to next 10, fact families, intuitive strategies)	Given an addition or subtraction problem, student uses strategies such as near doubles, adding 9, build to next 10, fact families, or intuitive strategies
Extending and applying addition and subtraction by using basic, derived, and intuitive strategies	Given a range of tasks (including problems with multi-digit numbers), student can complete them mentally, using the appropriate strategies and a clear understanding of key concepts

Fig. 3.4. A chart summarizing Clarke's (2001, p. 213) six key growth points in the domain of addition and subtraction strategies

These growth points align with our discussion of mathematics content in chapter 1. Diagnostic interviews are a means to assess students as they move forward on this trajectory. These tools can also explore the knowledge and thinking of students as they consider place-value concepts and addition and subtraction beyond the basic facts.

Diagnostic interviews also extend to more sophisticated understanding. For example, in a classic diagnostic interview, Ross (2002) explored students' understanding of the "ten as one" concept, which is central to place value. A version of the original task follows:

> Given a set of 36 counters, students are asked first to count them and then to write the corresponding number. When they have written 36, the interviewer circles the digit 6 and poses the question, "Does this part of your 36 have anything to do with how many counters there are?" Then the interviewer circles the digit 3 and repeats the question.

A diagnostic interview is meant to be a snapshot of students' understanding. The researchers evaluated the students' knowledge according to where their performances fell on a spectrum from merely selecting 6 counters and then 3 counters to a full understanding of the 3 as three tens and the 6 as six ones. For students to add or subtract skillfully, especially in cases involving regrouping, they need to demonstrate a full understanding of the value of each digit in the number. In Ross's study, many students did not fully understand the single-digit number in the tens column as 30. A student who lacks this understanding has little hope of carrying out addition and subtraction successfully and meaningfully.

Another diagnostic interview takes place when the teacher asks a student to complete an addition or subtraction and talk about what he or she is thinking in the process. For example, the teacher might say, "Add 99 and 145." A variety of performances might result. Some young students might try to align these numbers in columns (with some students misaligning the 99 over the 14 portion of 145), while older or more capable students might treat the 99 as 100 and then add it with a subtraction of 1 at the end to counteract their original jump to 100. This last example of evidence of student thinking clearly shows links to the study of derived facts as discussed previously in relation to Essential Understanding 2a.

One interesting aspect of the use of diagnostic interviews is the positive disposition toward mathematics that students seem to develop as they are asked to share and discuss their thinking individually with the teacher (Clarke 2001). As the students reflect on their thinking, their own awareness of their use of strategies is heightened. This gives them increased confidence to express themselves and believe that their effort and thinking are valued in the learning process. The interviews also provide teachers with precise evidence of the students' strengths and weaknesses, and this information can help in tailoring instruction and supporting all learners.

Essential ⬅
Understanding 2a

The commutative and associative properties for addition of whole numbers enable computations to be performed flexibly.

Is Subtraction Harder than Addition?

Researchers who have explored whether subtraction is more difficult than addition have said yes, it is (Baroody 1984; Fuson 1984). Consider the problem 9 + 3 = □. Counting on 3 more after 9 gives 10, 11, 12. This process corresponds to a visual display of 9 items and the action of introducing 3 more. Both of these methods align nicely with the concrete, semi-concrete (the extension of fingers or tally marks) and oral methods that students use. The case of subtraction is a bit different. Verbal solution methods in this case can be seen as "counting down to" a number, "counting down from" a number and "counting up to" a number (Fuson 1984). First, it should be noted that the usual concrete model for a problem such as 12 − 4 = □ would be a display of 12 items from which 4 would be removed, and then the remaining items would be recounted. This model does not match the three methods of "counting down to," "counting down from," or "counting up to." When students use a verbal solution approach, they must also decide whether to count down starting at the first number in the problem or at the number that is one less. So, for example, with 12 − 4 = □, students might verbalize "counting down to" as "12, 11, 10, 9, and then the answer is 8," though they might articulate "counting down from" as "11, 10, 9, 8, for an answer of 8." These distinctions are important since students can accidentally move between these approaches and become confused about the answer. This is the same confusion that emerges on initial use of the number line, when students confuse the unit of length with the points labeled by a number (and count the points instead of the spaces between them, a misunderstanding that Dougherty and colleagues [2010] address in their discussion of unit, count, measure, and number). In fact, Fuson suggests that students would benefit from the "counting up to" approach since they can apply it consistently and then "think addition" for subtraction situations. Also, a "counting up to" approach supports students who have issues with working memory or some ELLs who find the backward counting linguistically challenging. Furthermore, subtraction may be harder because students calculate differences from their knowledge of sums, making subtraction a two-part process for some students. For example, in trying to find 15 − 7 = □, the first part is identifying the corresponding addition problem of 7 + 8 = 15, and the second part is using that information to generate the answer of 8.

For an extended discussion of unit, count, measure, and number, see *Developing Essential Understanding of Number and Numeration for Teaching Mathematics in Prekindergarten– Grade 2* (Dougherty et al. 2010).

Conclusion

This chapter has examined the pedagogical considerations that emerge from thinking about the ways in which addition and subtraction are used to represent and solve many different kinds of problems. Becoming mathematically proficient in addition and subtraction (and developing proficiency in all five UCARE components) requires a learning environment that focuses on meaningful approaches to interpreting story problems, flexible ways to record equations, multiple ways to solve addition or subtraction problems, and numerous opportunities to connect concrete representations to abstract concepts. Furthermore, only through thoughtful conversations during mathematics instruction and by listening carefully for evidence of students' understanding can you diagnose your students' misunderstandings, identify their strengths, and help them move ahead.

Together, the three chapters in this book can serve as a resource for developing your mathematical understanding of addition and subtraction and helping you translate those understandings into engaging learning experiences for your students. First and second graders can often add quite well, using at least one algorithm or particular model. If you take away (no pun intended!) just one idea from this book, it should be that it is essential to engage students in using different models and a range of strategies (algorithms) for solving addition and subtraction problems. By doing so, you will position your students to develop a deep understanding of addition and subtraction, equipping them as completely as possible for success in later mathematical topics and a lifetime of competence in mathematics.

References

Barnett-Clarke, Carne, William Fisher, Rick Marks, and Sharon Ross. *Developing Essential Understanding of Rational Numbers for Teaching Mathematics in Grades 3–5*. Essential Understanding Series. Reston, Va.: National Council of Teachers of Mathematics, 2010.

Baroody, Arthur J. "Children's Difficulties in Subtraction: Some Causes and Questions." *Journal for Research in Mathematics Education* 15 (May 1984): 203–13.

Carpenter, Thomas P. "Learning to Add and Subtract: An Exercise in Problem Solving." In *Teaching and Learning Mathematical Problem Solving: Multiple Research Perspectives*, edited by Edward A. Silver, pp. 17–40. Hillsdale, N.J.: Lawrence Erlbaum Associates, 1985.

Carpenter, Thomas P., and James M. Moser. "The Development of Addition and Subtraction Concepts." In *Acquisition of Mathematics Concepts and Processes*, edited by Richard Lesh and Marsha Landau, pp. 7–44. New York: Academic Press, 1983.

Carpenter, Thomas P., Megan Loef Franke, and Linda Levi. *Thinking Mathematically: Integrating Arithmetic and Algebra in Elementary School*. Portsmouth, N.H.: Heinemann, 2003.

Clarke, Doug. "Understanding, Assessing and Developing Young Children's Mathematical Thinking: Research as a Powerful Tool for Professional Growth." In *Numeracy and Beyond: Proceedings of the 24th Annual Conference of the Mathematics Education Research Group of Australasia* (MERGA), vol. 1, edited by Janette Bobis, Bob Perry, and Michael Mitchelmore, pp. 9–26. Sydney: MERGA, 2001.

Clarke, Doug M., Anne Roche, and Annie Mitchell. "Ten Practical Tips for Making Fractions Come Alive and Make Sense." *Mathematics Teaching in the Middle School* 13 (March 2008): 373–80.

Common Core State Standards Initiative. *Common Core State Standards for Mathematics. Common Core State Standards (College- and Career-Readiness Standards and K–12 Standards in English Language Arts and Math)*. Washington, D.C.: National Governors Association Center for Best Practices and the Council of Chief State School Officers, 2010. http://www.corestandards.org.

Dougherty, Barbara J., Alfinio Flores, Everett Louis, and Catherine Sophian. *Developing Essential Understanding of Number and Numeration for Teaching Mathematics in Prekindergarten–Grade 2*. Essential Understanding Series. Reston, Va.: National Council of Teachers of Mathematics, 2010.

Fuchs, Lynn S., Douglas Fuchs, Caitlin Craddock, Kurstin N. Hollenbeck, and Carol L. Hamlett. "Effects of Small-Group Tutoring with and without Validated Classroom Instruction on At-Risk Students' Math Problem Solving: Are Two Tiers of Prevention Better than One?" *Journal of Educational Psychology* 100 (November 2008): 491–509.

Fuchs, Lynn S., Douglas Fuchs, Robin Finelli, Susan J. Courey, and Carol L. Hamlett. "Expanding Schema-Based Transfer Instruction to Help Third Graders Solve Real-Life Mathematical Problems." *American Educational Research Journal* 41 (Summer 2004): 419–45.

Fuson, Karen. "More Complexities in Subtraction." *Journal for Research in Mathematics Education* 15 (May 1984): 214–25.

–––. "Research on Whole Number Addition and Subtraction." In *Handbook of Research on Mathematics Teaching and Learning*, a project of the National Council of Teachers of Mathematics, edited by Douglas A. Grouws, pp. 243–75. New York: Macmillan, 1992.

Heddens, James. *Today's Mathematics: A Guide to Concepts and Methods in Elementary School Mathematics*. Chicago: Science Research Associates, 1964.

Kamii, Constance, and Ann Dominick. "The Harmful Effects of Algorithms in Grades 1–4." In *The Teaching and Learning of Algorithms in School Mathematics*, 1998 Yearbook of the National Council of Teachers of Mathematics (NCTM), edited by Lorna J. Morrow, pp. 130–40. Reston, Va.: NCTM, 1998.

Kenney, Joan M., Euthecia Hancewicz, Loretta Heuer, Diana Metsisto, and Cynthia L. Tuttle. *Literacy Strategies for Improving Mathematics Instruction*. Alexandria, Va.: Association for Supervision and Curriculum Development (ASCD), 2005.

Kilpatrick, Jeremy, Jane Swafford, and Bradford Findell, eds. *Adding It Up: Helping Children Learn Mathematics*. Washington, D.C.: National Academy Press, 2001.

Lamon, Susan J. *Teaching Fractions and Ratios for Understanding*. London and Mahwah, New Jersey: Lawrence Erlbaum Associates, 1999.

Leinhardt, Gaea, Ralph T. Putnam, Mary Kay Stein, and Juliet Baxter. "Where Subject Knowledge Matters." In *Advances in Research on Teaching*, vol. 2, edited by Jere Brophy, pp. 87–113. Greenwich, Conn.: JAI Press, 1991.

Lochhead, Jack, and Jose Mestre. "From Words to Algebra: Mending Misconceptions." In *The Ideas of Algebra, K–12*, 1988 Yearbook of the National Council of Teachers of Mathematics (NCTM), edited by Arthur F. Coxford, pp. 127–35. Reston, Va.: NCTM, 1988.

Ma, Liping. *Knowing and Teaching Elementary Mathematics: Teachers' Understanding of Fundamental Mathematics in China and the United States.* Mahwah, N.J.: Lawrence Erlbaum Associates, 1999.

Mack, Nancy K. "Confounding Whole-Number and Fraction Concepts When Building on Informal Knowledge." *Journal of Research in Mathematics Education* 26 (November 1995): 422–41.

McNeil, Nicole M., and Martha W. Alibali. "Knowledge Change as a Function of Mathematics Experience: All Contexts Are Not Created Equal." *Journal of Cognition and Development* 6 (January 2005): 285–306.

National Council of Teachers of Mathematics (NCTM). *Principles and Standards for School Mathematics.* Reston, Va.: NCTM, 2000.

———. *Curriculum Focal Points for Prekindergarten through Grade 8 Mathematics: A Quest for Coherence.* Reston, Va.: NCTM, 2006.

———. *Focus in High School Mathematics: Reasoning and Sense Making.* Reston, Va.: NCTM, 2009.

National Research Council. *Helping Children Learn Mathematics.* Mathematics Learning Study Committee, Jeremy Kilpatrick and Jane Swafford, eds. Center for Education, Division of Behavioral and Social Sciences and Education. Washington, D.C.: National Academy Press, 2002.

———. *Mathematics Learning in Early Childhood: Paths toward Excellence and Equity.* Committee on Early Childhood Mathematics, Christopher T. Cross, Taniesha A. Woods, and Heidi Schweingruber, eds. Center for Education, Division of Behavioral and Social Sciences and Education. Washington, D.C.: National Academies Press, 2009.

Otto, Albert Dean, Janet Caldwell, Cheryl Ann Lubinski, and Sarah Wallus Hancock. *Developing Essential Understanding of Multiplication and Division for Teaching Mathematics in Grades 3–5.* Essential Understanding Series. Reston, Va.: National Council of Teachers of Mathematics, 2011.

Piaget, Jean. *The Child's Conception of the World.* London: Routledge and Kegan Paul, 1928.

RAND Mathematics Study Panel. *Mathematical Proficiency for All Students: Toward a Strategic Research and Development Program in Mathematics Education.* Santa Monica, Calif.: RAND Corporation, 2003.

Ross, Sharon R. "Place Value: Problem Solving and Written Assessment." *Teaching Children Mathematics* 8 (March 2002): 419–25.

Ross, Susan C., and Mary Pratt-Cotter. "Subtraction from a Historical Perspective." *School Science and Mathematics* 99 (November 1999): 389–93.

Schifter, Deborah, Virginia Bastable, and Susan Jo Russell. *Building a System of Tens Casebook*. Developing Mathematical Ideas: Number and Operations, Part 1. Parsippany, N.J.: Dale Seymour Publications, 1999.

Seo, Kyoung-Hye, and Herbert P. Ginsburg. "'You've Got to Carefully Read the Math Sentence...': Classroom Context and Children's Interpretations of the Equals Sign." In *The Development of Arithmetic Concepts and Skills: Constructing Adaptive Expertise*, edited by Arthur J. Baroody and Ann Dowker, pp. 161–87. Mahwah, N.J.: Lawrence Erlbaum Associates, 2003.

Siebert, Daniel, and Nicole Gaskin. "Creating, Naming, and Justifying Fractions." *Teaching Children Mathematics* 12 (April 2006): 394–400.

Smith, John P., III. "Competent Reasoning with Rational Numbers." *Cognition and Instruction* 13 (March 1995): 3–50.

Starkey, Prentice, and Rochel Gelman. "The Development of Addition and Subtraction Abilities Prior to Formal Schooling in Arithmetic." In *Addition and Subtraction: A Cognitive Perspective*, edited by Thomas P. Carpenter, James M. Moser, and Thomas A. Romberg, pp. 99-116. Hillsdale, N.J.: Lawrence Erlbaum Associates, 1982.

Thompson, Ian. "Getting Your Head around Mental Calculation." In *Issues in Teaching Numeracy in Primary Schools*, edited by Ian Thompson, pp. 145–56. Buckingham, UK: Open University Press, 1999.

Van de Walle, John A., Karen S. Karp, Jennifer M. Bay-Williams. *Elementary and Middle School Mathematics: Teaching Developmentally*. 7th ed. New York: Allyn and Bacon, 2010.

Verschaffel, Lieven, Brian Greer, and Erik De Corte. "Whole Number Concepts and Operations." In *Second Handbook of Research on Mathematics Teaching and Learning*, edited by Frank K. Lester, pp. 557–628. Charlotte, N.C.: Information Age; Reston, Va.: National Council of Teachers of Mathematics, 2007.

Witzel, Bradley, Cecil D. Mercer, and David M. Miller. "Teaching Algebra to Students with Learning Difficulties: An Investigation of an Explicit Instruction Model." *Learning Disabilities Research and Practice* 18 (May 2003): 121-31.

Xin, Yan Ping, Asha K. Jitendra, and Andria Deatline-Buchman. "Effects of Mathematical Word Problem-Solving Instruction on Middle School Students with Learning Problems." *Journal of Special Education* 39 (Fall 2005): 181–92.

Yackel, Erna. "A Foundation for Algebraic Reasoning in the Early Grades." *Teaching Children Mathematics* 3 (February 1997): 276-80.